The Streetw

Fourth Edition

This down-to-earth guide to survival in the contracts jungle is enlightening reading for Specialist and Trade Contractors. It suits sole traders and those at a range of levels in firms of every size and specialisation operating in the often contentious construction and engineering industries, and particularly focuses on the relationships between the Specialist Contractor and the Main Contractor or the Client.

The fourth edition features all the old favourites such as payment, delay and disruption, extension of time and the all-important checklists and site records. But it has now been brought bang up to date to reflect the importance of the 2011 revisions to the Construction Act and the emergence of adjudication as the pre-eminent means of dispute resolution.

It provides clear insights and practical advice on avoiding contractual disasters through easy to follow routines. It is brief, crisp, easy to read, and occasionally light hearted. The text is supported by forms and checklists that have already made it onto site cabin walls around the country.

Barry Ashmore has overhauled this respected guide by drawing on his 48 years of experience, first as a Specialist Contractor and then as an adjudicator and consultant at the sharp end resolving disputes and solving contractual problems for Specialist Contractors.

The Streetwise Subbie

Fourth Edition

Barry Ashmore

Illustrations by Dave Eastbury

Routledge
Taylor & Francis Group

LONDON AND NEW YORK

First edition published by Butterworth-Heinemann in 1999

This edition published in 2018 by Routledge
2 Park Square, Milton Park, Abingdon, Oxon OX14 4RN
711 Third Avenue, New York, NY 10017

Routledge is an imprint of the Taylor & Francis Group, an informa business

© 2018 Taylor & Francis Group LLC

British Library Cataloguing in Publication Data
A catalogue record for this book is available from the British Library

Library of Congress Cataloging in Publication Data
Names: Ashmore, Barry J., 1954- author. | Russell, Jack, 1933 February 7- author.
Title: The streetwise subbie / Barry Ashmore.
Description: Fourth edition. | Boca Raton : Taylor & Francis, CRC Press, 2018. | Revised edition of: The streetwise subbie : how to survive the contracts jungle / Jack Russell ; illustrations by Dave Eastbury. 2nd ed. Oxford ; Boston : Butterworth-Heinemann, 2001. | Includes bibliographical references and index.
Identifiers: LCCN 2017061188 (print) | LCCN 2018000249 (ebook) | ISBN 9781351047197 (Adobe PDF) | ISBN 9781351047180 (ePub) | ISBN 9781351047173
(Mobipocket) | ISBN 9781138300163 (pbk.) | ISBN 9781138300156 (hardback) | ISBN 9781351047203 (ebook)
Subjects: LCSH: Subcontractors. | Construction industry--Management. | Construction industry--Subcontracting. | Small business--Management.
Classification: LCC HD9715.G7 (ebook) | LCC HD9715.G7 R87 2018 (print) | DDC
624.068/4--dc23
LC record available at https://lccn.loc.gov/2017061188

ISBN: 978-1-138-30015-6 (Hardback)
ISBN: 978-1-138-30016-3 (Paperback)
ISBN: 978-1-351-04720-3 (ebook)

Printed and bound by CPI Group (UK) Ltd, Croydon, CR0 4YY

Contents

Preface

As the late Jack Russell said in the preface to the Third Edition, 'this book is intended to serve as a practical guide to survival for trade and subcontractors of all kinds'. The original version of *The Streetwise Subbie* was published by Electrical Times in June 1996, since then it has been expanded from the 'little yellow book' into this practical guide, which has seen three previous editions.

In this fourth edition, I have taken the opportunity of amending and updating Jack's original material to reflect some of the more recent developments in our industry. This edition also deals with the new versions of the JCT and NEC subcontracts and the current working practices (some would say 'sharp practice') of some Contractors.

Old favourites include payments, delay and disruption, extension of time and the all-important checklists and site records. But the work now reflects the importance of the 2011 revisions to the Construction Act, and adjudication as the pre-eminent method of dispute resolution.

Finally, the opportunity has been taken to revise some of the existing chapters, to take account of my own personal experience of 20 years as a Specialist Contractor and 28 years as a professional Construction Contracts Consultant providing contractual advice to and resolving disputes for Specialist Contractors.

Jack's original concept was that a more systematic and 'streetwise' approach by Specialist Contractors would protect them from contractual and commercial disasters. This fourth edition is intended to build on that principle, and to provide a down to earth no-nonsense practical guide to survival in the contracts jungle.

It is intended for use as a working manual by every trade and Subcontractor, regardless of specialisation or size, and, most importantly, for general issue to all managerial, technical and supervisory personnel.

As Jack would say 'good luck, and be careful out there'.

Barry J Ashmore, DipLaw DipArb FCIArb MCMi

Note

Whilst the advice set out herein is given in good faith, it is just one practitioner's view. Readers should always take appropriate professional advice on their particular problems, and neither the author nor the publishers can accept any legal responsibility for the views expressed.

Author

 **Barry J Ashmore** is an established Contractual Consultant with 20 years' experience as a Specialist Contractor and 28 years in professional practice, advising Specialist and Trade Contractors. Formally trained in law, negotiation, mediation, adjudication and arbitration, his no-nonsense objective style and straightforward advice is trusted by Specialist Contractors throughout the United Kingdom.

He has set up and runs StreetwiseSubbie.com, the nationwide network of professionals and online advice service for Specialist Contractors.

Section 1

Worth the risk?

If you read nothing else in this book, please read this chapter. Of all the challenges that Specialist Subcontractors face, taking unnecessary risks must be the one that causes the most significant problems. Construction is a tricky balance of risk against rewards, so if you're frightened of risk, then you should find another way of earning a living. But that does not mean that risk is something to be ignored. So, my most important message is right here: please do not enter into any contract, or situation with your eyes wide shut!

Here are some ways of recognising and controlling the risks you are taking:

1. Ensure you know exactly who it is that you will be entering into contract with.
2. Check their financial credentials.
3. Identify the risks and responsibilities at tender stage.
4. If the risks are unacceptable, consider withdrawing.
5. Price for the risks.
6. Qualify or reduce risk by negotiation.
7. Ensure the contract incorporates what you have agreed.
8. Manage the risks.

Step 1 You may think you know who you are going to be working for but make certain. People move around and often take their Subcontractors with them. But the fact that you or a colleague might know someone at the company will count for nothing if their company fails. Even large national Contractors have subsidiaries or trading companies that are there to shield the parent if things go wrong financially.

Step 2 Check the company's financial status. Are they good for the money? If they fail **(and even the biggest can; think Carillion)**, and are not able to pay you, then all your hard work will have been for nothing, and it could cost you your business!

Step 3 Before you commit resources to the costly process of producing your tender, check out the terms and conditions that you are going to be asked to sign up for. Check for **onerous conditions**. These might include the following:

- Terms that nullify the benefits of the Construction Act
- Onerous amendments to Standard Form contracts
- Use of in-house forms of contract
- Excessive liquidated damages or unlimited damages
- Extended payment periods
- Non-payment for unfixed materials
- Conditions precedent for payment
- Excessive discount
- Extended retention periods
- Onerous set-off arrangements
- Onerous performance bonds and warranties
- Enforced acceleration without payment
- Lack of firm programme dates and periods
- Suicide terms such as 'to suit main contractor's progress'
- Excessive design or coordination responsibilities
- Unworkable protection obligations
- Responsibility for checking previous works of other trades

Step 4 You need to decide if the risks are bad enough to justify 'walking away' at this stage. This will obviously depend on a whole range of factors, including the current state of the market and your order book. But putting your whole business at risk just to fill a gap in workload could be the worst business decision you ever make!

Steps 5 Make a realistic appraisal of the risks identified from the enquiry documents and practical assessment as to 'what the market will stand' in terms of price and qualifications. Few Contractors will reject a favourable tender out of hand, even if there are some qualifications, and as price is almost always the key factor they will negotiate if the price is right. However, beware ruling your tender invalid in cases where qualifications are forbidden (e.g. local authorities and public utilities).

Steps 6 When it comes to agreeing to the terms of the contract, don't simply accept what the Contractor puts in front of you. Negotiate! Contrary to popular belief Contractors will negotiate about their terms.

Step 7 Once you have negotiated the best deal possible, make sure that what you have agreed on is properly incorporated into the contract, and don't start work unless you know and have confirmed exactly what basis you are starting work on.

Some Contractors are not averse to deliberately putting back in the contract that which you have negotiated out!

Step 8 You must manage the risk at every stage of the process. Here are a few ways in which you can manage risk:

- Use a RAID log to record all the risks from the outset (see Section 9 'What is normal').
- Give the site management a thorough 'team' briefing or 'workshop' as to the contents of the documents.
- Use the tender risk appraisal to instill awareness of the risks.
- Identifying these risks immediately they appear on the horizon.
- Use a system of site records and notices which seek to minimise and 'manage' the risks.
- Allocate sufficient and appropriate staff resources.
- Use a procedure for regular monitoring.

These procedures need to be operated as a matter of routine on every job. This can be done via standard checklists, linked in to the company's QA procedures (see the Appendices). One of the many benefits of this approach is that everyone is encouraged to feel 'part of the team'. If this approach is followed on all jobs, then financial disasters should become a thing of the past.

It's close to 30 years since Jack Russell first wrote *The Streetwise Subbie*. Since then much has changed, and in contractual matters mostly for the worse as far as Subcontractors are concerned. Back then most Contractors weren't out to screw their Subcontractors. Sadly nowadays, many can only survive by doing so!

Don't say you haven't been warned!

Identify and manage risk from the outset!

Beware letters of intent

The letter of intent is one of the most common sources of subsequent argument, or even disaster! Many such letters, when studied carefully, are often no more than an indication that some party is contemplating placing an order. If that party changes its mind, there is usually no legal comeback whatsoever .

If you are not careful, you could end up with all sorts of obligations, e.g. to commit to placing materials on order and organising resources, but with no entitlement to be paid for your trouble!

What is required, as an absolute minimum, is

1. **Instruction to proceed** and/or expend money on specific functions (e.g. 'Commence working drawings and procurement of quotations for specialist items')

2. **Terms of payment** for all works and/or services provided, including profit, as the work progresses and irrespective of whether or not the subcontract proceeds or formal contracts are entered into

3. **Confirmation of price**, and either no restriction or an appropriate restriction of the amount you will be able to recover

4. **Clearly defined scope of works**

5. **Confirmation of programme**, start date and periods

6. **Confirmation of agreed terms and conditions** – see 'Worth the risk?'

7. Indication of when **formal order or documents** will be forthcoming

8. **No onerous conditions**

The fact of the matter is that very few letters of intent comply with the above criteria. Regrettably, it is often the case that jobs are started (even major projects), on the basis of very vague letters which may well be contractually worthless. Even if they are indicative of the potential formation of a binding contract, there may well be glaring omissions regarding price, programme and/or other terms. These deficiencies are, in fact, the seeds of subsequent dispute. Indeed, it is fair to suggest that many disputes arise not from belligerence, but from uncertainty.

It is foolhardy to expend substantial resources on the basis of a 'letter of intent'. Even properly formed and well-drafted construction contracts are subject to interpretation, and they are complicated enough without the added burden of trying to establish whether or not there actually is a contract! The last place you want to be is in court trying to resolve a protracted and expensive dispute.

As a streetwise subbie, you should try to get all essential aspects (i.e. price, programme, subcontract conditions, etc.) agreed and confirmed in writing before you actually do any design, order any materials or commence work. If a letter of intent has to be the starting point, then at least make sure that it complies with the checklist set out above.

Better to risk upsetting the Client or Contractor at the outset than incurring substantial costs that you aren't going to get paid for!

Be very careful before starting work on a letter of intent!

Collateral warranties

It is almost inevitable that you will be asked to provide a collateral warranty. If these are included in or referred to in the enquiry documents, these need to be checked or qualified at the time of submitting your tender. Otherwise, you will be deemed to have accepted them as drafted.

Collateral warranties are agreements which are associated with another 'primary' contract. They provide for a duty of care to be extended by one of the contracting parties to a third party who is not party to the original contract. For most Subcontractors, this will mean that you are asked to provide a collateral warranty to the Employer who can be the Client or building owner, but very often they are also required for funders and tenants.

If you are asked to provide a warranty, you are being asked to warrant to a funder, tenant or purchaser that you have fulfilled your obligations under the subcontract. Collateral warranties often contain obligations such as using materials of an appropriate quality and carrying out work in a professional, workmanlike manner. But very often they will include multiple obligations.

The warranty enables the beneficiary to make a claim against you for losses which would not otherwise be recoverable. Conversely, they do not ordinarily provide you with any benefits whatsoever. Whilst the fundamental purpose of warranties is not in itself onerous, the terms that are often inserted into them are. Please do not assume that all warranties are the same. They most certainly are not, and some impose extremely onerous obligations!

Onerous non-standard clauses

Extending your obligations

One of the key objections to warranties is that they are often drafted without any reference to the obligations under the subcontract for the actual works. As a result, warranties often place greater or conflicting obligations upon the Subcontractor. For example, a warranty may impose a design responsibility upon you even if there is no such obligation in your subcontract!

First things first; before signing any warranty, you should ensure that there is a clause that restricts your obligations (and liability for breach thereof) under the warranty such that they are no greater than those under the subcontract. You certainly do not want to be accepting clauses like this one:

> Nothing in the Sub-Contractor's tender or in the Sub-Contract or the Main Contract or in any specification, drawing or other document put forward by the Sub-Contractor and no approval, consent or communication at any time given by or on behalf of the Employer or Main Contractor shall operate to exclude or limit the Sub-Contractor's liability for any breach of its obligations hereunder.

And, beware of seemingly innocuous clauses such as

> The Sub-Contract warrants and covenants with the Tenant/the Purchaser/ the Funder that he has observed and performed and will continue to observe and perform all the Terms and Obligations on his part to be observed and performed under the Subcontract, and any specification or obligations relating to the Subcontract works as are contained in the Main Contract.

Not only is this transferring all the terms of your subcontract into the warranty, but this is also making you responsible for anything that relates to your works that is tucked away in the Main Contract. For example,

- If you delay the works, you might be liable for the consequences of such delay to the beneficiary of the warranty.
- If the Contractor has not passed on details that affect your works, you will still be liable to the beneficiary, even if you have never seen them.

Essentially, your obligations in warranties should be confined to undertakings in respect of design, workmanship and materials.

Insurance

One of the most common obligations in warranties is the requirement to obtain insurance cover (usually professional indemnity cover) for anything up to 12 years and, in some cases, beyond that.

This requirement may apply irrespective of whether or not the necessary cover is available or will continue to be available. You should always check with your insurance company if you have any doubts about the contents of any warranty you are asked to sign.

Generally, the cover available will be limited to a maximum amount per claim and a maximum aggregate per year.

> The Sub-Contractor covenants with the Beneficiary that it has indemnity insurance cover with a limit of £3, 000, 000.00 for each and every claim, and he will henceforth retain such insurance cover for not less than that amount for a period of not less than 12 years from the date of practical completion of the works as certified under the terms of the Main Contract.

An alternative form of words requires you to obtain professional indemnity cover for so long as it is available at commercially reasonable rates. It is impossible to guarantee that the required cover will be available over the period demanded, and there is certainly no definition of what is 'commercially reasonable'. What may be commercially reasonable for some might seem extortionate to you!

Fitness for purpose obligations

Under the standard warranties, the Sub-Contractor warrants that he has exercised all reasonable skill and care in the design of the subcontract works. If you check out the Standard Form Contracts such as JCT DBSub/C 2016 Design and Build Sub-Contract, the design obligation is expressed as follows:

> Insofar as the design of the Sub-Contractor's Designed works is comprised in the Sub-Contractor's Proposals... the Sub-Contractor shall in respect of any inadequacy in such design have the like liability to the Contractor, whether under statute or otherwise, as would an architect, or other appropriate professional designer who holds himself out as competent to take on work for such design and who, acting independently under a separate contract with the Contractor, has supplied such design for or in connection with works to be carried out and completed by a building contractor who is not the supplier of the design.

This form of wording limits your liability to reasonable skill and care, and you should ensure that the standard of design required by warranties is no more than that of reasonable skill and care. Sometimes, slight variations on this theme are to be found such as the phrase reasonable skill, care and diligence. Diligence means taking care and giving proper attention to the matter in hand. And its inclusion is unlikely to add or detract anything from the foregoing words.

The main cause for concern are those clauses that seek to include the highest standard of legal liability which is that of fitness for purpose:

> The Sub-Contractor hereby covenants to design the Sub-Contract Works so that the design shall comply fully with all the requirements of the

drawings and specifications attached hereto and so that upon completion the Works shall operate and so shall also be suitable in every respect for the purpose for which they are intended.

Please do not be confused by this, or if you are please take professional advice, because 'fitness for purpose' is an extremely onerous liability that has serious consequences!

Probably the most serious consequence is that your insurance company will not provide cover in respect of fitness for purpose obligations, and you will have paid your very expensive premium for nothing!

Assignment

Under English law, rights and obligations under contracts are only enforceable between the parties of the contract. This is known as the privity of contract rule. By way of exception, the law allows a party to assign or transfer his rights and benefits under a contract to a third party.

Only the benefits of the warranties can be transferred or assigned to subsequent owners, tenants, etc. Warranties often provide for rights of unlimited assignment without the consent of the Subcontractor.

If you agree to unlimited assignment, it may be extremely difficult or impossible to obtain the necessary professional indemnity cover.

As far as the beneficiary is concerned, the right to assign warranties is one of the most important clauses in the warranty. However, a sensible limit on the number of times a warranty can be assigned must be negotiated.

insurance companies are generally unhappy if more than two assignments are stipulated.

Limits on your right to suspend and/or determine

It is very common to find restrictions on your right to suspend and/or determine your obligations under the subcontract.

Do not accept them! Such a right is particularly important where non-payment is concerned. Watch out for clauses such as

> The Sub-Contractor covenants and undertakes with the Beneficiary that the Sub-Contractor will not exercise any right which he may now or at any time have under the Sub-Contract or otherwise to determine the Sub-Contract or discontinue the performance of any of its obligations in respect of the Development in whole or in part without first giving to the Company not less than 28 days' notice of the Sub-Contractor's intentions specifying the grounds for the proposed suspension or determination.

You can rest assured that in this case, the Employer, who insisted upon the above require-ment, will not be so keen to ensure that you get paid during the 28-day notice period!

If you do accept this kind of provision, you destroy your most effective weapon in the fight to get paid. That weapon being your right to suspension!

Novation – 'Step-in rights'

The requirement to give notice before suspending or determining the subcontract is generally imposed to give the beneficiary, such as the Employer or a funder, an oppor-tunity to rescue the project if there are problems with the Main Contract such as the insolvency of the Contractor.

This may require the novation of the subcontract. This means that the beneficiary will want to step into the 'shoes' of the Contractor.

You must insist that the warranty expressly provides that if the beneficiary wants to do this, he will assume all the obligations owed to you the Subcontractor, includ-ing making payment of all outstanding sums due to you. This arrangement can only be achieved by agreement of the three parties involved – beneficiary, Contractor and Subcontractor.

Occasionally, clauses dealing with novation will restrict the liability of the benefi-ciary in respect of those obligations which have remained outstanding at the date of the novation agreement.

> In the event that the employment of the Contractor is determined for any reason whatsoever, the Subcontractor, shall carry out and complete its ob-ligations under this Agreement and shall enter into an Agreement within 14 days of the date of such determination novating the Sub-Contract to the Employer and the Subcontractor shall not thereafter unreasonably withhold his consent for further novation of the Sub-Contract by the Employer...

This clause in the warranty will continue in force even after the determination of your employment, and you will be required to agree to the Employer taking over from the Contractor and will have no right of objection!

More importantly, there is no requirement for the beneficiary to pay, and you will have no right to payment of the monies owed to you by the failed Contractor!

> **Collateral warranties are yet another contract to look out for!**

Make sure you check that order

No doubt you will have worked hard to get that new order. It means valuable workload, turnover and employment for your operatives. But when you have finished celebrating, it is vital to check that the value, scope, programme and conditions are those on which the tender was based and/or subsequent agreements reached. As boring a task as this may seem, it is a vital part of the process of entering into the contract. If you don't understand any of the small print, then please take professional advice. It doesn't have to be expensive, and paying a small fee now could save you £ thousands in the long run.

Clients and Contractors often introduce additional terms within their order. Many will try to introduce their own onerous terms even though standard terms have been agreed. In the 'Third Edition', Jack said that in his experience over some 25 years, the majority of orders are not satisfactory and need to be clarified and amended before proceeding further. Let me add my 28 years of professional experience to that and say that he was absolutely bang on!

You need to be aware that refusal to sign and return acknowledgement slips, or not signing and returning the contract does NOT mean that the terms haven't been agreed or won't apply! Any 'action' taken by you after the order has been received can be construed as 'acceptance by conduct'.

Most of these problems would not arise if Standard Form contracts were used, as recommended in the Streetwisesubbie.com Fair Treatment Charter, and by all the major trade and professional bodies. In the majority of cases, there is one reason, and one reason only why Standard Form contracts are not used, and that is the Standard Forms such as the JCT and NEC contracts are considered by the Contractor to be 'too fair' to the Subcontractor!

Watch out for onerous terms such as

- Extended payment periods
- Extended 'fixed price' periods
- 'Pay when and if certified' or other such variants – these are not enforceable if the Construction Act applies, but if it doesn't they are!

Recent case law is a reminder that the devil is in the details!

In a judgment issued on 3 August 2017, the Supreme Court has overruled the Court of Appeal decision in a recent case[*] and restored the TCC first instance finding.

Despite potentially conflicting obligations to exercise 'due care and professional skill' and compliance with international standards contained within the Main Contract terms, the contractor was liable to comply with a fitness for purpose obligation **contained within a technical requirement schedule.**

This decision will have consequences for the interpretation of construction contracts which incorporate technical requirements or specification documents. It is also a timely reminder that dual obligations (such as reasonable skill and care and fitness for purpose warranties) and **onerous duties may be contained in often overlooked schedules or documents referred to in the contract provisions.**

---

[*] MT Højgaard A/S v E.ON Climate and Renewables UK Robbin Rigg East Limited and Anor [2017] UKSC 59.

- No payment for unfixed materials, on- or off-site
- Compliance without payment or time – with the most onerous obligations in the case of mistakes or ambiguities in the information supplied to you
- Discount that is now rarely if ever linked to prompt payment
- Excessive retention percentages and/or periods
- Onerous 'set-off clauses'
- Onerous performance bonds and warranties
- Acceleration clauses without entitlement to reimbursement
- Entitlements restricted to the 'benefits under the Main Contract'
- Main Contractor's right to vary the subcontract programme and/or period
- Suicide phrases like 'To suit our programme'
- Unworkable and/or excessive 'design' and/or 'coordination' responsibilities
- Excessive 'protection' clauses
- Responsibility for previous works by other trades

- Client's 'milestone dates' for access and fit-outs without release of obligations
- Excessive rate of liquidated damages

My advice is that you should always agree terms BEFORE you do anything that constitutes acceptance. If you are foolhardy enough to start work without having agreed terms, it is vital that you make your position quite clear in writing, stating very carefully those matters with which you disagree. You must emphasise your rejection of unacceptable terms.

But, the bottom line is that starting work without agreeing terms puts you at risk, since you have no agreed contract to rely on!

Check the details before you enter into that contract!

Section 2

Programme and progress

Beware 'starting the clock'

Even if you haven't yet signed the contract, starting work on-site, doing design work or even ordering materials could be construed as acceptance by conduct. Alternatively, the contract will have been agreed and signed. In either case, any of these actions 'starts the clock ticking' and the agreed period begins to elapse. At the end of that period, if your works are incomplete, and in the absence of a formal extension of time, you are in breach of contract and the Contractor's or Client's remedy for that breach is damages.

Regrettably, in my experience the standard of management by the Contractor's team has declined over the last 10 years, as many Contractors have shed their most experienced staff. As workload has picked up the Contractors have brought in inexperienced personnel, or as is often the case freelance staff. On a great many projects, this leads to problems, including programming the works, and in some cases I have seen, the programming was shambolic.

Of course, the Contractor won't accept any of these responsibilities, and therefore if you are persuaded or bullied into starting work on an 'unready' site, it will put you

---

Measure of damages in contract

Damages are awarded for breach of contract. Generally, the purpose of an award of damages for breach of contract is to compensate the injured party. The general rule is that damages are meant to place the claimant in the same position as if the breach had not occurred.

Damages for breach of contract are subject to the principles of remoteness causation and mitigation. In addition, it is commonplace that there are liquidated damages in the contract between the Contractor and the Employer.

As a Specialist Subcontractor you are at risk of incurring a 'thick sandwich' of damages if you do not finish your works by the agreed date, and do not have an extension of time.

That 'sandwich' comprises liquidated damages, the Contractor's own costs and any other Subcontractors who have been delayed.

at great disadvantage and risk! Having commenced, you will find your team 'working out of sync' with the other trades. Then, when the Contractor finally 'gets his act together', you will be accused of causing delay!

You must visit and inspect the site well in advance of the planned start date. If the site clearly will not be 'ready', then record the specific reasons why by writing to the Contractor. Of course, you can discuss this with the Contractor first-hand, but you must still ensure that you follow any procedure that is set out in the contract for notifying delay. You should collect as much information about the matters that are preventing you from starting the works as possible, including annotated photographs.

Be aware that 'uneconomic working conditions' are not a valid reason for not starting the works if the agreed commencement date has arrived. That is a matter which has to be dealt with by way of subsequent extension of time and/or loss and expense.

If a deferred start date or revised periods are mutually agreed, then record the agreement in writing and confirm your entitlement to a revised completion date (based upon the original agreed period commencing at the new start date). But again, you must ensure that what has been agreed is formalised in accordance with the subcontract conditions.

If, despite your protests about the state of readiness, the Contractor still gives you formal 'notice to commence', then (providing the site is safe for working) you probably have to comply. Again, it all depends on the precise subcontract terms. If you then find yourself 'delayed from the start' (i.e. by lack of access and/or building readiness, etc.), you will have to give a formal delay notice at 'day one' and continue to do so for as long as the delays continue.

If you end up sending 'two men and a dog' to do a couple of days work on an unready site, you will still have 'started the clock ticking'. So, do you refuse to co-operate? Most certainly not! What you do is to cover yourself with a nice polite letter which confirms that you are doing so against your wishes and in accordance with the specific instructions of the Contractor, pointing out that this will cause uneconomical working and that you will incur costs which you will be seeking to recover.

Remember all the time that you MUST comply with the terms of the Subcontract in respect to the notices that are required under the contract. See 'don't be shy – get noticed'.

In the third edition, Jack warned about treating what he said as being too pessimistic? I reckon that almost **30 years on, following the above advice is even more critical**.

If you don't believe it can happen to you, then ignore my advice. But don't be surprised when those damages are knocked off your final payment!

> **Confirm your commencement date, and record all delays in writing.**

A programme checklist

Programmes present many pitfalls for the unwary Subcontractor. So what should the 'streetwise' subbie bear in mind?

First and foremost, the Contractor's programme might be part of the contract documents, but under many of the Standard Forms of Contract, e.g. JCT, it does not define the Contractor's obligations in respect of time. And, in turn, you should be wary about entering into a subcontract where your obligations as to time are defined by reference to a programme rather than specific dates or periods.

The situation is quite different under the NEC forms of contract and this is one of the key differences as between the JCT and NEC forms. See 'The JCT and NEC Contracts'.

The completion date and any stage or sectional completion dates are enforceable and failure of the Contractor to meet these dates may lead to a claim by the Client for liquidated damages. Contracts will generally require that the Contractor progresses the works regularly and diligently and failure of the Contractor to meet the dates on the master programme might be evidence that this is not the case.

All of this and more can be and usually is stepped down into your subcontract with the Contractor, so here are a few suggestions about dealing with programmes:

Watch out for onerous provisions such as this

2.4.3 All dates for the subcontract works are abstracted from the Contractor's current programme and are subject to revision as the contract proceeds, the subcontractors programme of works will require him to comply specifically with the phasing requirements of the Main programme (or revisions thereto) of works and the subcontractor will be required to complete all aspects of his works, including commissioning where applicable to ensure no delays are incurred to the completion.

2.4.4 The issue of a revised programme by the Contractor shall not constitute an extension of time in accordance with Clause 2.5.

---

1. Agree a reasonably detailed programme for your works at the outset, and have it incorporated by specific reference in the subcontract order or agreement.

2. If it is impossible to get it in the formal agreement, then at the very least try to agree a subcontract programme right at the outset, and use it as your 'baseline'.

3. Watch out for and object to 'suicide' terms such as 'works to proceed in accordance with the Contractor's requirements' and 'to suit the progress of the Main Contractor'.

4. Try to obtain a copy of the Contractor's programme for subsequent monitoring.

5. Keep all programmes in a safe place, and print off electronic copies to keep on file and issue your site team with copies.

6. Ensure all programme negotiations and agreements are confirmed in writing at the time.

7. Ensure all programmes, both issued and received, are referenced, date stamped and covered by a letter that clarifies their status.

8. Maintain a programme register recording dates and identity of all programmes both issued and received.

9. Again, keep revisions in a safe place and issue the site with copies.

10. Promptly examine the implications of the Contractor's revised programmes and report to him accordingly and/or give him appropriate notices of delay as required by the contract.

11. Remember that whilst actual progress is usually monitored against the current revision, your 'contractual performance' (i.e. extension of time, exposure to damages, etc.) is measured against the actual dates for completion in the contract.

12. Keep an eye on the Contractor's progress, and monitor delays to other trades that have significant influence on your works. Where necessary, you must give appropriate notices of delay as required by the contract.

13. Bear in mind that a definite agreed programme imposes equally definite obligations on you as regards progress and performance, failure to achieve which might lead to financial penalties – you can't have your cake and eat it!

14. Don't be put off doing what needs to be done when it needs to be done, and don't be conned or persuaded by the Contractor not to do what needs to be done because it is 'not in the spirit of the contract' or because 'you're being contractual'.

Jack said don't allow your activities to drift into 'contractual limbo', and what he meant by that was that if you fail to take the necessary action, because you're too busy getting the job done, or any other reason, you will find yourself in a position that is very difficult to recover from. And it could cost you substantial sums of money or even (in extreme cases that I have seen more than once) your entire company!

What's that you say? You do all these things already, as a matter of course? Congratulations – you're unique.

Make sure you understand your obligations as to time from the outset.

Don't be shy – get noticed!

Most standard forms of subcontract require you to give notice of delay and/or extra cost in such terms as 'forthwith', 'as soon as reasonably apparent', etc. Indeed even if there are no such terms it is good practice to do so, and if the Contractor's own terms apply, they will almost always have very stringent and/or onerous notice requirements.

Either way, and never mind the finer nuances, it is quite simple. The party you are in contract with, e.g. the Contractor, is entitled to know, at all times, about anything that affects the timely completion and end cost of the project. If you do not issue notices, you are in breach of contract, and as a matter of legal principle you are not entitled to benefit from your own breach. So be warned!

In the third edition, Jack said, 'It has become "the norm" for the subbie to suffer delays and financial loss due to the defaults of other parties'. And decades on it has almost become the only way that some Contractors can make money, i.e. by taking money off their Subcontractors. In my experience, the Contractor almost welcomes the project being delayed. The Contractor then finds any reason he can to blame the Employer or the Client and persuades them not to deduct liquidated damages in exchange for the Contractor not claiming loss and/or expense for the extension of time he says he is entitled to.

This then leaves the Contractor free to blame all of his Subcontractors for the delay. The Contractor then sets about hacking big lumps off each of the Subcontractor's accounts for the delay he says the subbie has caused.

A typical scenario can see your works delayed by lack of access, poor building progress and lack of progress by other key trades; delayed information; and excessive and/or late variations. These and any number of other reasons result in 'piecemeal' working and delay and disruption to your works.

Many such situations go unrecognised (and unnotified) until it is too late. The danger is that disruption and 'hand to mouth' working become part of daily life on-site and are accepted as 'normal'. A good rule of thumb is to stop and ask 'did the estimator price the job to be done this way?' In almost every case, the true answer would be 'no way!'

You must be ready and able to identify these circumstances as soon as they appear on the horizon and to notify the Contractor in writing forthwith. Do not be put off by accusations that you are 'getting contractual' or any other BS that the Contractor wants to conjure up! You have a contractual obligation to serve notice as soon as you foresee delay, and/or disruption, and the terms of the contract will almost always require you to notify delay within a matter of days or you will lose your right to claim the costs and the loss and/or expense that will inevitably arise.

Make no mistake, the Contractor will be the first to rely on 'lack of proper notice' as a basis for refuting your claims and bashing you with 'set off' charges for alleged delays. There is no need to antagonise the Contractor, even if he reacts badly. Indeed it is good practice to stay on as friendly terms as possible. Call him up or go and see him. Explain that the contract he has put in place requires you to give notice and that you will lose your entitlements if you don't. But don't be put off.

Having given notice, be ready and willing to discuss the situation in a constructive and friendly manner (without committing yourself to additional expenditure). Submit daily records of any labour or plant involvement. Submit as much detail of 'cause and effect', i.e. the relationship between events or things, where one is the result of the other.

A typical example: the 'cause' is you are prevented from completing the installation of light fittings because the suspended ceiling in one of the rooms had not been installed. The 'effect' is that your works are delayed until the ceiling is installed, and abortive costs arise from the disruption. Such costs would include the time taken to restock the light fittings back into the site storage or back to base; the cost of removing the plant you were using, e.g. hop ups or other; and the time for your operative to tell the foreman, the foreman to redeploy the operative and tell the Contract Manager (CM), the CM to tell the Contractor, etc.

This is a combination of action and reaction, and in some way, shape or form, the details need to be captured at the time because it's a hopeless exercise if left too late! Provide an estimate and details of likely loss and expense as soon as possible.

You must tell the Contractor immediately and confirm it in writing. Give him the chance of doing something about it. Remember, an early notice and outline of possible costs stands far more chance of getting sympathetic treatment than some 'ripping yarn' submitted long after the job is completed.

No notice means you are in breach of contract and/or no extension of time. No extension of time means you are at the mercy of the Contractor when he's looking for someone to take money off!

Make sure you notify the Contractor in writing about all problems.

Some key thoughts on notices

The importance of giving notice when problems arise cannot be over emphasised, so in addition to 'Don't be shy-get noticed!' here are some key thoughts about notices:

- Before the works commence, read the contract very carefully to ensure that you understand how notices are to be served. Establish whether your notice requirements give rise to any conditions precedent by the use of words such as 'condition', 'subject to' or 'condition precedent'.

- Amendments to the standard forms or the Contractor's bespoke terms will inevitably shorten/specify the time by which notices are to be issued, which is likely to require notices to be given at the earliest possible date. You should therefore check your contract carefully for any such amendments so that you are clear as to what is required.

- Before you serve a notice, be clear on the form the notice is to take. Notices will usually have to be given in writing. If no form is prescribed, you should describe the event or the circumstances relied on in detail, and make it clear on the face of the notice the claim to which the notice is intended to relate under the contract. The notice must be recognisable as a 'claim'.

- To avoid any arguments that your notice has not been properly served, try and ensure that your notice is served in the correct way, e.g. on the right office, and on the correct person, e.g. on a director, the project manager or other senior personnel of the recipient as might be required by the contract.

- If you do not serve notices correctly, and on time, and your contract imposes a condition precedent in relation to the time for service of notices, you risk losing entitlements to additional time or money to which you might otherwise be entitled. This is the case even if the notice itself was valid, and was served by the correct method on the correct person, so always proceed with care when issuing notices.

- If you are claiming an extension of time, the earliest you can serve your notice of claim is the earliest date on which it becomes clear that the works are delayed and will impact completion. To be on the safe side, make sure you notify as soon as it becomes likely that the works will be delayed.

The process of giving notice is crucial to the triggering of the contract mechanisms that allow you to pursue additional contractual entitlements. Most importantly extension of time and loss and/or expense. It is on the face of it a simple process, but it is one which a great many Subcontractors fail to get right. Either because they don't issue notices at all, or they don't issue contract compliant notices.

Please make sure you issue contract compliant notices. Failure to do so could be very costly indeed!

Recognise the importance of notices.

Why work for free?

Subcontractors spend much of their time 'working for free'!

That might seem crazy, and you wouldn't set out to do that would you? Of course not. But that is what happens in practice when you fail to recover your legitimate entitlements for work properly done, additional works, variations and reimbursement for delays.

Most of this loss can be avoided if certain basic routines are consistently maintained, and monitored, from day one to final handover. The following routines should be operated in a spirit of co-operation, and willingness to look for solutions to problems. But that does not mean being naive:

1. Confirm start and completion dates for the job and for each section or area.

2. Record date and origin of all programmes and revisions.

3. Notify all delays to the Contractor/Client immediately they become apparent.

4. Notify the effect of any delays upon the completion date.

5. Recognise and notify variations and claim situations ASAP and submit details

6. Be proactive in recommending solutions – but not at your expense!

7. Apply for extension of time as soon as any delay appears likely.

8. Make sure all instructions are in writing and don't work without instructions

9. Submit dayworks for signature at the time.

10. Keep good records of everything e.g. names, activity worked on and key events.

11. Be prompt in applying for payment, with detailed evaluations and forecasts.

12. Keep the pressure on for payment – priority goes to the man who makes most noise!

Act promptly or you could end up working for nothing!

Site records and survival

In the third edition, Jack said, 'If I had to choose, I would always prefer a set of good site records than the services of an expensive lawyer or claims consultant'. And I wholeheartedly agree, albeit that as a 'claims consultant' I have had some fairly good results using poor records. But it's hard work and costs you far more!

Keeping good records during the job costs time and money. However, it works out a lot cheaper than a major legal dispute at the end. The truth is that good records can make the vital difference between survival and disaster. With records and notices, you can prove your case for extension of time and additional costs. Without them, you will not have a prayer! So what are the essential records for survival?

Even on smaller jobs there should be a site diary, emails, site delay reports, confirmation of oral instructions, drawing register, daywork sheets and time allocation sheets. Some examples have been included in Appendix 4.

Larger, or potentially troubled jobs will require a higher degree of sophistication (e.g. programme/progress reports, technical queries, programme receipt/issue register and cause/effect records). Photographic evidence can be dynamite, but don't forget to number and date each photograph, and add the location and significance. Remember you are trying to show what has or hasn't been done, and why that affects you.

Site records should include names and numbers of supervision, labour and plant, Specialist Subcontractors; summary of progress and notable events in the day; principal delays and reasons for same; and notes on the Contractor's key progress or lack of it (e.g. roofing, weatherproofing, internal walls and ceilings).

The records should be regularly reviewed and actioned by the project team. Why should a site supervisor put a high priority on keeping records if the powers that be never show interest, or never use them? This is particularly relevant to troubled contracts, where the sheer tempo of events makes it harder to find time for record-keeping.

When the time comes to prove your entitlement, Clients and Contractors will demand detailed 'cause and effect' of all delays and additional costs. This means recording delays and disruptions in detail, and showing the effect upon programme and completion whilst it is there to be seen. If this is left to the end, it can be a nightmare of a task. Meanwhile, the 'guilty parties' have fled the scene! Having kept the records, don't keep them a secret. Submit them at the time, while the situation is 'hot'.

Some Subcontractors seem to think that disclosing records weakens their hand, or that they will somehow be able to ambush the Contractor with them later on. This is a major misconception.

When you start putting records forward, don't be put off by how the Contractor reacts. I recall advising one Client to do exactly what I am advocating here, and the Contractor saying, 'hang on, what are you doing?', to which my Client was able to reply, 'exactly what your contract says I must do!'

Finally, it is vital for management to review the situation on a daily, weekly or monthly basis depending on the circumstances. The objective is to ensure that you comply with your obligations under the contract and if necessary, request an extension of time and/or notify loss and expense.

The idea is to (a) to persuade the Contractor to recognise/address the current problems, (b) to protect you from damages and set off and (c) to preserve your entitlement to recover additional costs.

No records – no extension of time and no payment.

Records and codes

On most jobs, I expect to find all manner of documentation, confirmation of instructions, technical queries, delay notices and correspondence files. These records provide the database when challenged to prove our entitlements to be paid for variations, secure extensions of time, recover loss and expense or fight off an attempted contra charge from the Contractor.

This can mean trawling through 20 lever arch files at the end of a large job, abstracting individual references under specific headings and generally trying to piece together what has gone on. This process can take months, and it is one of the costliest phases of the battle to be paid what's rightfully yours.

You can greatly improve your chances, and save a lot of money, by utilising the technology that exists to support quick and easy analysis. In the third edition, Jack suggested 'adopting a "coded" approach to records, from the outset of every job' and to 'draw up a simple coding list, (e.g. "B" for basement) and a sub code (e.g. programme activity number). This list is then issued to all concerned, with brief instructions. Each individual is then required to state the code and subcode when issuing any site record or correspondence.'

Whilst things have moved on considerably in terms of the technology that exists, the basic systematic approach that Jack advocated all those years ago is just as valid today as it was back then.

Key points

A well-developed electronic and paper filing system is a crucial element of good construction project management.

Not only must the filing system keep track of all sorts of events, activities, locations, responsibilities, deliverables and deadlines, but it must also facilitate document retrieval from a variety of locations, and easy analysis. For this reason, you should adopt a standard file identification and coding system. A coded system makes file management significantly easier and facilitates analysis and communication between interested parties.

If this coded approach is adopted, it is a clerical exercise to input the information into a computerised database or spreadsheet as the job proceeds. It is then a simple matter to sort the data into codes and subcodes in date order. At the touch of a button we can see, for any given area or activity, how many queries have been raised or un-answered, variations received, delay notices issued, etc. That is pretty useful when discussing matters with the Contractor.

This approach involves only a little more time on the part of the issuer and costs very little to input and monitor. There are great time savings for all concerned during the job. Proving cause and effect becomes routine. Claims and variations are highlighted as the job proceeds, and are much quicker and cheaper to prepare.

So let's start coding with our next job.

Use technology to save time and money.

Site records checklist

Good site records are essential in order to record the history of the job, get paid for variations and delays, and secure extensions of time. Make no mistake records are fundamental to securing your entitlement in the event of a dispute and especially so if the matter has to be resolved by way of adjudication, arbitration or litigation. The site records are without doubt a key source of evidence.

In a perfect world, our site or project team should be given some training, or at least some guidance, as to what the company wish to see in the site records. However, it is a fact of life that for some reason, we assume that our team automatically know what to record and how to record it. This is a potentially costly misconception, when you consider that the money is made or lost at the work face.

Ideally, the site records should be kept in a standard format. Some examples can be found in Appendix 4. Please don't just issue a blank A4 diary from a well-known chemist, or allow the supervisor to do their own thing. And please make sure it's all backed up on your office system. I have seen all the records disappear on the supervisors laptop on more than one occasion!

Whether in preprinted or electronic format, there are certain key essentials that should be included in any site records.

Below is a checklist. Pin it on the cabin wall. Look at it every day, and ensure all your records capture the following information:

1. Date
2. Weather
3. Visitors
4. Labour details; numbers, names, locations
5. Subcontractors on site (identity, etc. as point four above)
6. Progress details
7. Area completions and handovers
8. Access, builder's works requirements
9. Delays due to same

10. Outstanding information requirements
11. Delays due to same
12. Overall building progress
13. Variations instructions received
14. Key events (e.g. Client's visits, meetings and major instructions)
15. Health and safety, accidents

Ensure that good records are kept for the entire duration of the job, even when the supervisor is on holiday or off sick, and keep it going right up to the very end of the job.

Ensure all site records are monitored as the job proceeds. Why should a supervisor bother to keep records if nobody else shows an interest?

Lastly, make sure that any hard copy site records are returned to the main office when the site is cleared.

Good site records are worth their weight in gold.

What makes a good delay notice?

First and foremost, it is essential that you comply with the requirements of the contract. If you are required to send the notice on pink paper in a green envelope tied with string, and on the day the problem occurs for it to be valid, then that is what you must do! Trust me, I have seen notice requirements that are almost that ridiculous.

It is optimistic to expect that the site engineer and/or supervisor will always have at their elbow the full subcontract conditions, etc. So, in order to facilitate the issue of contractually correct notices, it is imperative that everyone involved in the project knows exactly what is required. It is equally important to take prompt action rather than to spend time creating a legalistic document.

The requirements of most subcontract conditions are very similar. There will be a when, a what, a who and a mechanism. It boils down to this – you must notify the Contractor/Client immediately there is a 'problem', partly because, in theory at least, the Contractor will take action to resolve the problem so that your works can proceed without delay. And partly because nobody likes 'nasty surprises'. Furthermore, he is entitled to be told the exact nature of the problem, how it came about, the immediate effect on your programme and progress, the likely effect on overall completion and any cost implications.

And do NOT think, 'the Contractor knows he hasn't put the roof on and the rain is coming in'. Of course he does, but you MUST STILL give notice!

Give notice in writing forthwith, and in accordance with the contract, for each individual delaying or disrupting event as soon as it becomes apparent.

The basics of a good notice are listed below:

1. State the area and location of the problem (e.g. Level 1 Restaurant).

2. State the exact circumstances causing the delay or disturbance (i.e. identify the precise 'cause' of the problem).

3. Identify the relevant provisions of the contract under which you are giving notice. The JCT contracts refer to 'Relevant events' (e.g. Architect's instructions, late information, delay caused by the Employer, lack of access and/or building progress). Whereas the NEC contracts refer to 'Compensation Events'.

4. Give the expected effect on your programme/progress (i.e. state which subcontract activities are affected and how).

5. State what action you require from other parties in order to avoid or reduce the effect of the delay (e.g. remove scaffolding, pump out water and provide information).

6. If your overall completion date is likely to be affected, give an estimate of the delay and the revised completion date.

7. Give notice of any cost effects, with details if appropriate.

8. Update the notice as necessary, if the delay continues.

9. Don't forget to record when the delay has ceased and the final effect.

If early warning is given, it may be possible to nip the problem 'in the bud'. That will be better for everyone. Indeed, you should always take a proactive approach and be ready to suggest a way forward. The bottom line is that you will protect yourself from possible set-off charges and liquidated damages when the project overruns. And you will protect your right to extension of time and the recovery of prolongation and/or disruption costs (e.g. additional labour costs, site preliminaries, staff, cabins and plant and overheads).

Good notices are essential to protect your interests.

Get the picture?

Nowadays everyone has a very powerful camera in their pocket in their mobile phone, but site progress photographs are often a source of disappointment. When viewed a year later, they usually tell the observer very little about the actual state of affairs on the job. This is a great pity, because a good collection of site photographs can be of considerable benefit in demonstrating the reasons for delayed completion. As usual, the rules are simple enough and yet seldom followed:

1. Don't just concentrate on the works that you have installed to date. Try to capture the state of the environment and the resultant problems (e.g. lack of building structures to be provided by others and restrictions of access).

2. Capture evidence of your operatives working in adverse circumstances, and the surrounding difficulties.

3. Make sure each photograph is dated and individually referenced.

4. Maintain a register on a standard format, each photograph identified with its unique reference number, other information boxes for location, activity involved, special notes regarding difficulties and/or surrounding conditions.

5. Use the photographs to cross reference with other internal records (e.g. progress reports).

6. Above all, keep the photographs properly filed and backed up, and make sure that people know they exist.

Good progress photographs are a valuable source of information.

Just a minute

'Don't worry – it's all covered in the minutes'. As Jack said in earlier editions, his 'poor old heart used to sink' when he heard those words, and with good reason. Relying on the minutes of meetings to prove your entitlements is a recipe for disaster! So, on a practical note, what are the pitfalls and tips regarding site minutes?

1. Minutes of meetings are usually written by the other party, e.g. the Contractor, and will be slanted, cleverly or otherwise, in their favour. This 'slanting' ranges from the omission of important statements made by you during the meeting to the more subtle black art of using just the right words.

2. In order to be streetwise you must check all minutes immediately you receive them, and immediately write back asking to have them corrected with any points of omission and/or disagreement. These corrections must then be included in the minutes of the next meeting, as the very first item.

3. Read them as though you were an 'outsider' who did not attend the meeting. Then you will often notice the clever bits, which appear to represent your comments but are worded so as to leave a different, retrospective interpretation, should the Contractor find it necessary later on. If you stated quite simply that you have been prevented from achieving progress for 10 weeks in a particular area due to lack of progress by others on internal block walls, then that is what you are entitled to see. A clever Contractor might report the item as 'Joe Bloggs Ltd said that there had been initial delays with block walls, and Ace Builders Ltd said that this was now back on programme'. Not the same really, is it?

4. Watch out for the Contractor who issues the minutes of the previous month's meeting on the morning of the current meeting. This is done so that you miss out on your chance of correcting them. Don't

fall for it. Register your concern at the current meeting, and insist that the minutes be issued within say 3 days of any meeting.

5. When you respond in a meeting, choose your words carefully. If you are being delayed, and you are in no doubt as to the causes (e.g. specific building or information delays or variations) then don't be 'mealy mouthed' about it, as is so often the case. The streetwise subbie must state his opinion clearly and ensure that it goes into the minutes as stated. If the contractor disagrees, that is a separate matter, and ideally the facts can be established and these can then be minuted.

6. If there are problems due to a fundamental underlying situation (i.e. extreme delay in completion of the roof, and subsequent effect on weatherproofing, etc.), then try to see that this is reflected in the minutes. Otherwise, it will just read like a series of minor trivia, which fails to convey the real situation on site. If for a long period the site resembled a bombsite, it is convenient months after completion, for all this to have faded from people's memories, and very real problems to be trivialised.

7. Above all, don't rely on minutes as a substitute for proper notices and good records. Whilst minutes can be very helpful evidence in substantiating case for delay, loss and expense, etc., they are no more than that.

Don't rely on meeting minutes, and always check them for accuracy.

The twilight zone

How sad it is that despite all the talk, and all the initiatives and reports that have been produced about construction over the years, that the same old problems still exist.

In previous editions, Jack wrote about a typical scenario: Site not really ready at the outset, but the subbie provided two men and a dog to show willing. Struggling for months to find work faces, until the Contractor finally gets his act together. Then tremendous pressure placed on the subbie to flood the site with labour at his own cost.

Meanwhile, there is a flow of late 'variations' as the Architect or Employer contributes to the chaos by introducing lots of late instructions. The Subcontractors are then bullied into unpaid acceleration, but the job runs over by months anyway, due to the Contractor's incompetence and the mass of additional works that should have been in the project design in the first place.

The result is often a financial disaster for the subbie by way of additional costs, unpaid variations and spurious contra charges from the Contractor for alleged delays.

Such is the situation that Jack described, and all these years later I'm still seeing on a regular basis. If anything, it's actually getting worse!

And so the accusations of delay start flying around, huge chunks of money are knocked off the subbie's account and the battle is on to prove your entitlement to time and money.

This is the 'twilight zone' a place where you definitely don't want to be, because it's going to cost you money, it's going to distract you from doing what you do best, and it's going to be very stressful. The moral is clear, take the right action at the right time to avoid the 'twilight zone'!

> **Take the right action at the right time to avoid the 'twilight zone'.**

Keep an eye on the Main Contract programme!

The site records kept by the typical subbie often ignore the 'big picture'. For example, there will be a reference to individual delays awaiting specific workfaces, other trades, etc. However, there is often no mention of the overall delay in construction of the building envelope itself.

Yet this can often be the dominant factor governing progress on the site. If there is substantial delay in construction of the external block walls, carcassing and covering of the roof, installation and glazing of window frames, etc., and if all this is exacerbated by lack of temporary weatherproofing, there will be a massive impact upon the subbie's progress. And yet, the subbie's records might tell us nothing of this. As a result, the execution of the subcontract installation is being carried out in radically different conditions to those which could reasonably have been contemplated at tender stage, or by reference to the agreed programme. For instance, it has now become common to see half-finished buildings with walls painted and carpets down in some areas, yet in other areas finishing trades are climbing over heaps of rubble as they endeavour to carry out second fix, not to mention carrying out expensive final fix, installations. All this taking place in a building that is woefully unready and insecure. No wonder it's so hard to make any money.

So, I recommend that the streetwise subbie tries to get hold of a Main Contract programme, preferably at the outset. That might be a tall order as some Contractors are reluctant to release such information. But as they say; 'if you don't ask you don't get'. And if you ask in a 'this will greatly assist us in efficiently programming our works' kind of way, then you are legitimately entitled to confirm their decision back to them if they won't provide it. After all the inference will be that by refusing to provide the programme, the Contractor is by implication preventing you from 'efficiently programming' your works.

Another way to monitor the Contractor's overall progress is to identify key stages of completion which will impact on your works, e.g. 'building weathertight' or 'blockwork complete to office area'.

Either way, the point is that what you want is a mechanism by which you can monitor and record the Contractor's overall progress. And you can then subtly or otherwise include reference to the Contractor's progress in your own progress reports and notices. Exactly how this is done will depend on the circumstances of your particular project.

Admittedly, this is not a way to win the Contractor's 'subbie of the year' award. Nevertheless, a record of the overall building delays can be one of the best defences against any attempt by the Contractor to blame the subbie for delayed completion of the project at a later stage. And believe you me an awful lot of Contractors will blame the subbies for causing delay, even if they are completely innocent!

The problem here is that if the Contractor has caused the problem, the buck will stop with the Contractor, since the Employer or Client will not be interested in circumstances that derive from default on the part of the Contractor. However, the subbie's contractual relationship is with the Contractor, and if the worst comes to the worst, then the Contractor will be looking for someone to blame.

Don't want to bother with any of this. Don't want to upset the Contractor? OK, so what's another fifty grand down the chute? You will suffer, but you will make the Contractor's commercial director a very happy man as he collects his bonus!

It's very often the case that the Contractor has an obligation to the Employer to provide his overall programme. That is certainly the case in NEC/4 (previously NEC/3) contracts (I would recommend reading Section 8). And as it's almost inevitable that the Contractor's programme will change, the Contractor's revised programmes issued throughout the job will be of particular value to the subbie.

So, my advice is don't ignore the overall picture and focus solely on your own activities. Because, this is in fact, a golden opportunity to monitor the overall progress of the project, not by subjective opinion, but by the slippage evidenced in the Contractor's revised programmes.

This can be readily achieved by a simple spreadsheet, which highlights the slippage occurring in the key structural activities and identifies the effects upon the subbie's own progress. A much-simplified example is given below, and as what you will actually have to produce is likely to be much more involved. So, if you don't have the relevant skills, or time in house, then you should seek professional assistance.

A spreadsheet of this kind is very good evidence to demonstrate the fundamental reasons for the delay to your works, and to secure extension of time and financial compensation. However, it is no use carrying out this exercise after the job has finally ground to a finish. The subbie has a positive duty to notify all delays in writing as they become apparent. The streetwise subbie must respond to the Contractor's revised programme and/or delayed construction progress at the time in order to protect your entitlements.

Keep a close eye on the Contractor's overall progress.

Activity ref.	Activity	Original completion date	Revised completion date	Slippage in weeks	Effect on electrical progress
20	External block walls	20 Apr 2016	17 Jul 2016	12.5	Delay to first fix
26	Roof carcassing	24 May 2016	20 Aug 2016	12.5	Delay to first fix
32	Roof felting	10 Jun 2016	3 Sep 2016	12	Delay to first fix
36	Roof covering	14 Jul 2016	7 Oct 2016	12	Delay to first fix
40	Stairs pc conc	31 May 2016	27 Aug 2016	12.5	Delay to first fix
46	Install windows	28 Jun 2016	22 Sep 2016	12	Delay to second fix
53	Window glazing	24 Aug 2016	7 Oct 2016	6	Delay to second fix
54	Watertight date	25 Aug 2016	8 Oct 2016	6	Delay to second fix
75	Statutory electric mains	7 Oct 2016	16 Jan 2017	15	Delay to power on
87	External doors	10 Nov 2016	17 Jan 2017	9.5	Delay to final fix
90	Internal doors	27 Nov 2016	20 Jan 2017	8	Delay to final fix
101	External works	14 Dec 2016	14 Feb 2017	9	Delay to external lighting
104	Gate house	14 Dec 2016	15 Mar 2017	13	Delay to CCTV
105	Hand over	20 Dec 2016	22 Mar 2017	13	Delay to practical completion

Section 3

Personal factors

Little tin gods

Get your site team on board from day one

Your site team are 'at the sharp end'. They are out there on-site every day, up to their ears in 'muck and bullets', the visible face of your company. Never forget, the site is where the money is spent and the financial returns are generated. It is also the place where most of the problems arise and have to be resolved. And it is one of the key things that sets our industry apart from say manufacturing, where the physical work is taking place in a controlled environment, with the management team close at hand.

My journey in construction started 'on the tools' aged 16, and back then there was none of this 'can't go on-site until your 18' nonsense! You were straight in at the deep end, and I even volunteered to work 'out of town', and lodged away from home, rather than spend what was then an obligatory period in the stores counting out nuts and bolts. Then, when I had qualified as an electrician, I was offered a job in the office and my 'managerial' career was born. As part of that first spell 'in the office', I was briefly back out on-site as a 'supervisor', and in a similar fashion to Jack's recollections in the third edition, the experience also taught me how hard it is to get tradesmen to do as they should and to get the best out of them.

Over the years, I too acquired a healthy respect for site teams and supervisors and the tough job they have to do. A good supervisor was often the key difference between the job making money or losing money. And yet back then it was almost unheard of to give supervisors, or indeed the project managers (or as we were called 'contracts engineers'), any formal training. Thankfully, certain aspects are better some 40 years later, and I have had the pleasure of providing 'commercial awareness' training for a number of firms including their supervisors. But, in my own personal view, there is still much more to be done, and the industry should be tackling this issue at a grass roots level to ensure that Subcontractor's management, site teams and supervisors get the training and support they deserve.

We should take positive steps to get the site team on board at the beginning of every job, involving them in the pre-planning, discussing the key objectives and strategy. Having done that, we should keep them involved at every stage, including regular team meetings at which everybody plays a part. Such a policy avoids the typical 'I thought Fred was handling it' syndrome. Why exclude from the discussion those who really know what's going on out there at the workface? And yet this is often the case. Hardly the way to inspire enthusiasm.

As for site records, the diary, delay notices and photographs, do we explain at the start of the job exactly why these records are important for our protection, and the possible disaster which may follow if we neglect them? Do we regularly monitor the records? Even some major companies are neglecting the basics, and no one is reviewing what the site team is actually doing. If that is the case, then it should come as no surprise if the site time neglects the collection of good, accurate and useful records. Why should they bother when nobody else shows interest? It is absolutely essential to check all site records on a regular basis. This can be done using the checklists (Appendices 1 and 2), and/or by including the check within your quality assurance procedures.

Once managers take an active interest in the records the site team are producing, there is invariably an improvement in their quality and usefulness. This 'Hawthorne effect' was the subject of a study at the Hawthorne Works (a Western Electric factory outside Chicago), where workers' productivity seemed to improve when changes were made, and slumped when the study ended. It was suggested that the productivity gain occurred as a result of the motivational effect on the workers of the interest being shown in them.

So, could you be showing more interest in what your site team is doing, or do more to get them on board from day one?

The site team are often the key to success – get them on board at day one.

Adverse reactions

I have emphasised the importance of submitting prompt notices, with backup details, as soon as you see that you are likely to be delayed, disrupted and/or incur additional costs. The subcontract conditions oblige you to do so. Furthermore, it is only fair to all concerned that problems should be identified early on, so that joint action can be discussed to overcome them.

However, it is a fact of life that many Clients and Contractors will react adversely. Over 20 years ago, in the very first edition of this book, Jack referred to the subbie being accused of 'getting contractual'. Despite all the well-meaning reports and expressions of collaboration that we have seen in the intervening period, it is incredibly frustrating and sad to see the same line being trotted out by Contractors today! There is definitely a 'bully boy' culture in certain parts of the industry, and perhaps this tells us a lot about some of the people who work in it.

In my tongue-in-cheek publication *The Little Book of Crap Advice for Specialist Contractors*, I recount the following true story about a £500,000 problem which I helped the subbie to resolve:

> Once upon a time there was a nice blue-chip oil company who employed a Specialist Subcontractor to install lots and lots of steelwork. Things didn't quite go according to plan, so the Subcontractor sent a polite letter to the nice blue-chip oil company.
>
> 'Oh dear, please don't do that' said the nice oil company's Project Manager, 'It isn't really in the spirit of things, and you're being contractual.'
>
> 'Oh, I'm so sorry' said the Subcontractor, 'I do apologise, I won't do it again'. And nor did he. He was a brave little soldier and didn't say anything else for almost a year despite lots and lots of problems. Then one day his wallet was empty...
>
> 'Awfully sorry to trouble you', he said to the nice oil company (who had lots and lots of money – after all it was an oil company). 'I am rather short of money, in fact, about £500,000 short actually.'
>
> 'Oh dear', said the nice oil company, 'Oh dear..., Oh dear...You see, I'm afraid we can't pay you'. 'Why's that', asked the Subcontractor politely.

'Well, it's quite simple really' said the nice oil company. 'Our terms and conditions say quite clearly that you should send us notices when things aren't quite going according to plan...

You didn't send us any notices, so I'm afraid we can't pay you.' 'Now bugger off'.

So, next time someone suggests you are 'being contractual' just remember the true story above, and that these are the same people who will happily deduct 'set-off' from monies due, pointing to the absence of proper delay notice as justification. If you feel the need to do so, here are a couple of examples of how you might respond to such comments:

1. Our company believes that we have an important duty to keep you informed as problems arise, so that we can tackle them together at the time and avoid end-of-job disputes.

2. We are only doing what your subcontract conditions require of us, and we would be in 'breach' if we did not.

3. These notices are part of our QA procedure to ensure that we comply with the subcontract conditions.

Stay calm and professional when faced with adverse reactions.

Little tin gods

It is not unusual for the Client or Employer to be represented on-site by a high-profile personality who gradually assumes the status, to all intents and purposes, of a 'little tin god'. This is particularly the case where the subbie is working 'in the Client's backyard' (e.g. existing works, factories and hospitals). Indeed, in many such cases, security and/ or safety make it necessary that someone like a plant manager has ultimate control over movements of all in his domain.

In consequence, the perceived extent of that individual's contractual authority can expand far beyond anything in the documents, if indeed he has any such powers at all. All too often, the overenthusiastic subbie falls into this trap.

Unfortunately, the process of recovering entitlements, both as to time and money, at the end of such a job, can be long and expensive. At best, the subbie may be told that the 'little tin god' had no authority. Alternatively, 'LTG' may have been promoted to a new plant in Outer Mongolia or Scunthorpe, where communications are notoriously difficult.

To a lesser extent, the same situation can apply to such people as consulting engineers, clerks of works, architects and the growing army of 'coordinators'.

Please check what your contract says. Do not simply take instructions from 'LTG' or any of the other various parties you will come into contact with. And if the contract isn't clear, then make sure you seek clarification where necessary, and get it recorded in writing at the outset. A little extra trouble at the start can save a financial nightmare later on!

Check out the authority of those giving instructions.

Section 4

Acceleration

Best endeavours or acceleration?

Today's projects tend to follow a familiar pattern. The building is already behind programme, but theSubcontractor is pressured, against his better judgement, to make a start. Despite the lack of weatherproofing and work faces, it is always a case of 'jam tomorrow'. And so, the poor old subbie gallantly presses on, often on a 'seek and find' basis, whilst his original period gently ticks away. Two-thirds through the original period, he has installed one-third of his work scope. Sound familiar?

But now the real fun starts. The Contractor begins to 'get his act together'. Other trades such as ceiling fixers and plasterers are bullied into working weekends. A delayed bulge of work faces becomes available. Our subbie is told that the project completion date must be met. He is told to increase his labour force, and to work weekends, to make up 'his' delay, or else! He is reminded that the contract says he has to use 'best endeavours'.

If the subbie hasn't served his delay notices, he will be in very big trouble. Weekend working and doubling of labour strength can cost a fortune.

Whilst 'best endeavours' certainly implies an obligation to rearrange activities and to make all reasonable efforts to prevent or reduce the delay, and possibly incur some costs as a result, most textbooks agree that it should not involve substantial expenditure.

What is happening is, in fact, 'hidden acceleration'. It is up to you to recognise it and address it as such. Those subcontract conditions that envisage 'acceleration' usually provide for prior agreement as to reimbursement. Unless, of course, you have been foolhardy enough to agree to an 'accelerate for free' type clause. In many other cases, no provisions exist, and the matter of 'acceleration' stands to be discussed, and an agreement reached.

The Contractor should be politely reminded as to the origins of the delay and of the Subcontractor's entitlements to extension of time. The extent and nature of possible 'special measures' should be discussed and agreed at this stage, as should the vital matter of payment. Is it to be a 'lump sum' or 'milestone payments'? Perhaps a 'formula' basis of premium time plus a percentage is to cover overheads and profit, additional supervision, loss of productivity due to the 'fatigue' factor, etc.

However you tackle it, I would recommend that you take professional advice, and get it agreed in writing, before you start taking action and incurring costs! Please do not accept vague promises to 'reimburse all reasonable costs', for such assurances are virtually meaningless, as many subbies of all sizes have found to their cost.

If you allow yourself to be sucked in to the overall shambles without first reaching a firm 'agreement', then you had better have deep pockets or an understanding bank manager!

Recognise and deal with 'hidden acceleration'.

Acceleration – the true costs

We have already seen how easy it is to be sucked into accelerating for free by a clever Contractor under the guise of 'best endeavours'. We have emphasised the importance of getting formal agreement to reimbursement first. In the third edition, Jack referred to a comprehensive report that is available from the Chartered Institute of Building, Ascot, Berkshire, called 'Effects of accelerated working, delays and disruption on labour productivity'. It is an old report, but so far as I am aware its findings are still as valid today as they were back then, and if you have £20 or so to spare, you could do worse than buy a copy.

Here are some of the authors' nightmare conclusions:

1. There is a 5% loss of productivity for every 5 hours increase in the working week. So, a 60-hour week can result in a 20% reduction in productivity. The longer the period of overtime, the greater the loss.

2. Attempts to avoid extended hours by use of a 'shift' system can result in wasteful 'overlap' time for both men and supervisors.

3. Use of 'weekend guests' involves premium payments and lack of job knowledge, also absence of 'project commitment'.

4. High levels of pay associated with accelerated working may actually cause an increase in absenteeism.

5. A 50% increase in the workforce can cause a 15% reduction in productivity. A 100% increase therefore can cause a 30% reduction.

6. As density of labour increases, so does congestion of work faces, particularly if other trades are suffering from similar conditions. Coordination and control become much harder.

7. A cut of 50% in available work faces may cause a productivity loss of 20%.

8. Interruptions lasting more than half an hour can cause a productivity loss of 20% for the rest of that day.

The above conclusions are supported by a profusion of detailed graphic evidence. Frightening, isn't it? All the more so, when you realise that some 20 years on from when Jack referred to this report, the project management skills of some Contractors have actually got worse. What that means in practice is that the programming and coordination of all the various trades by the Contractor is shambolic. And to cover their tracks, the Contractor has to bully the Subcontractors into accelerated working.

Ignore this report at your peril!

Don't be bullied into accelerating into a financial disaster.

The alligator problem

There is a very old saying that 'When you're up to the backside in alligators, it's hard to remember you originally went in to drain the swamp'. The anonymous author thus aptly described the eternal dilemma of the typical 'man at the sharp end'. Ravaged by the dramas and crises of the day, harassed by impatient Clients and bullying Contractors, it is a tough job to keep the show going from one day to the next.

It is even more difficult, in such a typical scenario, for the unfortunate subbie to retain any overall perspective of the job. As a result, the accumulation of job changes and building delays can gradually change the nature of the work (i.e. 'draining of the swamp'). However, the extent and negative consequences of the change often goes unnoticed by the subbie. The Contractor is usually too busy doing his Rambo impression to know, less still care, how this is impacting on productivity. It is a small wonder that so many subbies go to the wall.

So, what can you do? One technique is to set a fixed day each month for a 'stand back day'. On this day, you lock yourself away for a couple of hours, you forget the immediate crises and you ask 'Is this what I started out to do?' Very often, the answer will be 'No way!' For instance, the fact that you are still on site after 50 weeks should be compared with an original 35-week period. Your current labour force of 30 operatives should be compared with an original plan for 10 men. And did the documents tell you there would be a daily deluge of 'ad hoc' variations? Get the idea?

Having taken time to stand back from your daily battle with the 'alligators', you can see that you are no longer 'draining a swamp' but dredging the Zambezi. Having identified your situation, you must now act! This requires some courage, since no one else will want to know. As to actions required, terms which spring to mind include 'delay notice', 'extension of time request', 'claim for loss and expense', 'repricing of variations and affected contract work', etc.

All too often, events go by unrecognised for lack of a regular and methodical review process. Then, when the review takes place, we are already dealing with a crisis. We are 'reacting' to events, not 'controlling' them! It's a bit like waiting until we have toothache before looking for a dentist. It's a lot smarter to go along for a regular 'check-up', and nip problems in the bud. To exercise control over events, there is no better aid than a 'checklist'. In the appendices to this book, you will find two such documents.

The Streetwise Subbie's site checklist (Appendix 3)

This is a simple list of basic reminders for display on the site cabin wall. It is grouped under the five main headings of programme, progress, information, claims and records. It can also be used as a weekly or monthly report. The intention is to ensure that you maintain basic procedures on site for ensuring contractual protection.

Monthly check-up (Appendix 4)

This is a much more comprehensive document providing a review of every key contractual aspect. It is intended as a report to be submitted to management on a regular monthly basis. Despite the apparent complexity, it should take no more than 15 minutes to systematically tick through the questions and even less to look at the answers.

May I suggest that you have your first 'stand back day' next week on current jobs, and as a matter of routine on all future jobs.

Is what is actually happening on site what you set out to do?

Section 5

Cash flow and variations

Are your systems user friendly?

Cash flow – everybody's problem

I have attended many meetings in my time, both as a Subcontractor and as a Consultant advising the Subcontractor. Payment and cash flow are always high up the agenda, and it is often somewhat surprising to see how badly the whole process of actually getting paid is managed.

What is equally surprising is that responsibility for getting paid is often laid at just one person's door, and that this person's actions are critical to the company's abundance or, more likely, lack of funds. Unfortunately, such a view, whilst commonly held, is an oversimplification of the real position.

In reality, 'cash flow' is the responsibility of every individual in the organisation, and success or failure starts from day one.

Key point: even profitable businesses go bust

It seems oxymoronic that a business making plenty of sales can go under. However, making sales and having money in the bank is not the same thing. You could be signing plenty of deals for handsome amounts, but until you collect the money, you can't spend it. Consequently, if the money isn't flowing into the business to satisfy creditors and suppliers when it's needed, then trouble beckons. In such circumstances, even profitable businesses can go bust.

Some of the main reasons why a successful business can run into trouble are as follows:

- Focusing on profit instead of cash flow
- Paying suppliers too quickly
- Poor control of the application/payment process
- Failing to understand the payment process
- Not submitting/agreeing variations
- Extending time on-site without claiming additional costs
- Extending credit to the wrong customers

If

- Work is acquired on extended payment terms,
- The contract says the Contractor can set off contra charges he 'might' incur,
- Orders are placed with suppliers and Subcontractors on unfavourable terms,
- Labour resources are expended without regard to the tendered allowances,
- Variations are accepted and implemented without any real attempt to incorporate them in the current interim application,
- Delays and disruptions go unrecorded and unnotified,
- 'Special measures' such as weekend working and imported labour are undertaken without any prior agreement,
- Essential site records such as site diaries and time allocations are neglected,
- Office records and systems such as wages and invoices are not readily accessible and user-friendly,

then the 'cash flow' will be bad!

In other words, all the various functions need to be geared from the outset towards 'good cash flow'. This starts with the tendering to order process, where every attempt should be made to agree sensible payment terms and to reject onerous terms.

In my experience, a Contractor will often be willing to negotiate, providing your tender price is sufficiently attractive, or there is some other compelling reason to use you.

And having negotiated, the subcontract agreement or order should be carefully vetted to ensure that it faithfully represents what has been agreed.

The programming and resourcing of the job should be done with one eye on the tender allowances. Labour costs should not just be allowed to happen, they need to be carefully controlled.

Just because it's in the tender doesn't mean you have to spend it. Your responsibility is to meet your obligations under the contract, not to spend what the estimator has allowed.

Site staff should be encouraged to notify the Contractor immediately if there is any kind of delay or disruption, and to keep detailed records.

The Contractor should be kept fully informed at all times regarding anything which may affect completion or costs, and variations and claims should be kept fully up to date, and agreed on a rolling basis, and incorporated into your applications.

You need to understand exactly how the payment process works and escalate any non-payment immediately. If you can, suspend the performance of your obligations if you don't get paid, but take advice before you do so. Never let matters drag on.

Utopia? Maybe. But, it is clear from the above that good cash flow should be the job of everyone in the firm. Only when we face up to this basic fact can we stand a chance of having good cash flow. And you don't want to pay the ultimate price of bad cash flow!

Don't neglect cash flow – even profitable businesses go bust.

Payment problems

Many payment problems are self-inflicted! The time to deal with payment is right at the outset, when vetting the initial enquiry from the Client or Contractor. Here is a useful checklist of things to look out for, which assumes that the Construction Act applies:

1. What is the interval between commencement and first payment?
2. When is the 'Due Date'?
3. What is the duration from Due Date to 'Final Date' for payment?
4. Is payment linked to Main Contract certification?
5. What is the timing for the Payment and Pay Less Notices?
6. Who values the works?
7. How often?
8. Are unfixed materials on-site and off-site to be paid for?
9. When is 'final payment' due?
10. What is the retention percentage?
11. When is it reduced, and is it reduced by half?
12. When is final retention release date?
13. How long is the defects liability period?
14. Is there a 'set-off' clause and is it onerous?
15. Make sure any discount is 'cash discount' or is otherwise linked to prompt payment.
16. Do the terms actually comply with the Construction Act?

If all the above matters are checked and unacceptable terms addressed at tender stage, then many 'payment problems' would not occur!

Every application should include everything that you are entitled to, and that includes work carried out on variations. Make sure you keep your variation pricing and submissions up to date. Very few domestic subcontracts require approval of the costs by the Client's quantity surveyor. Yet many of the delays in interim and final payment are due to the Contractor saying: 'The Client's quantity surveyor has not had time to look at your variations.' Be prepared to say: 'So what? My contract is with you!'

If there isn't already a schedule of all the dates relevant to getting paid, then establish the valuation dates, Due Dates, and Final Dates along with the dates when you should receive the Contractor's Payment and Pay Less Notices. Plot all this on a spreadsheet. You should then monitor the whole process as it proceeds, and take action if the dates are not adhered to instead of waiting for non-receipt of the cheque to trigger off your actions.

Have your accounts call up to check that a payment is on the system and that the Contractor's accounts department has received the necessary paperwork to release your payment. If there is a problem, be polite but be persistent. If you must, be a nuisance. Insist on the Contractor giving you proper notice of payments due and build-ups of what's included.

Similarly, with any sums withheld from payments due, ensure that you know why your valuation has been reduced, and why your variations have been slashed! If necessary, consider suspension of your works (but make sure you are entitled to do so and give the necessary period of notice in writing).

If you don't get paid, don't prevaricate, escalate! The squeaky wheel gets the grease!

Avoid payment problems before they happen-but if you have to, you must take action!

Don't get angry – get paid!

If you don't get paid your business fails. It's that simple! Ask any insolvency practitioner and they will confirm that lots of otherwise profitable businesses fail because they don't collect what's owed to them.

Of course, actually getting paid is sometimes easier said than done, it can be a very emotive, frustrating, and darn right stressful issue. And 'how can we get paid' is certainly one of the questions that I am most commonly asked, and the one which I have personally been advising Specialist Contractors about for the last 28 years.

So, my advice is don't get angry, get paid, by understanding the following key points:

1. The payment provisions of a contract should be clearly established and fully understood.

   It's a sad fact that a great many Specialist Contractors enter into contracts without really knowing what the exact payment terms are.

   Most standard form contracts, e.g. JCT and NEC/3 have provisions entitling you to interim payments, with clear rules set out as to how these interim payments should be calculated and when they should be made. Unfortunately, a great many Contractors seek to amend these standard form contracts, or simply draft their own terms.

   You may have been told that; 'the payment terms are 45 days', or something equally vague, and accepted that at face value. But when you don't actually get paid 'on time' and seek assistance to resolve the matter, you may find that the contract you have unwittingly agreed to actually contains all sorts of complex terms and unfathomable terminology that conspires against you.

   Of course, the contract might be formed by exchange of correspondence, and may even be silent about payment, in which case the Construction Act may come to the rescue, if it applies. Prior to the Construction Act there was no automatic right to interim payments, and there were no payment notices, no suspension, no adjudication, etc., so despite the Act making contracts a bit more complicated, it is generally good news for Specialist Contractors.

2. Be aware that the Construction Act (the Act) does not apply to all contracts. So, first things first please make sure you read and understand the contract

and whether or not the Act applies. There are certain types of project to which the Act does not apply, and one such situation that a good many readers will encounter is where the contract is with a 'residential occupier', the home-owner to you and me. Other situations where the Act does not apply include supply only contracts, process industry and power generation sites. A full list of the types of project to which it does not apply are set out in the Act as follows:

a. *Drilling for, or extraction of, oil or natural gas.*

b. *Extraction (whether by underground or surface working) of minerals, tunneling or boring, or construction of underground works, for this purpose.*

c. *Assembly, installation or demolition of plant or machinery, or erection or demolition of steelwork for the purposes of supporting or providing access to plant or machinery, on a site where the primary activity is*

   i. *Nuclear processing, power generation, or water or effluent treatment, or*

   ii. *The production, transmission, processing or bulk storage (other than warehousing) of chemicals, pharmaceuticals, oil, gas, steel or food and drink.*

d. *Manufacture or delivery to site of*

   i. *Building or engineering components or equipment,*

   ii. *Materials, plant or machinery or*

   iii. *Components for systems of heating, lighting, air-conditioning, ventilation, power supply, drainage, sanitation, water supply or fire protection, or for security or communications systems, except under a contract which also provides for their installation.*

e. *The making, installation and repair of artistic works, being sculptures, murals and other works, which are wholly artistic in nature.*

3. Even if the Act applies the payment terms can still be onerous.

Unfortunately, even if the Act applies some Contractors will draft their terms, or amend Standard Forms, so that the payment terms comply with the Act but are still nonetheless onerous. The most obvious way Contractors abuse the act is to extend the payment periods. The Act itself does not prescribe any timescales in respect of the Due Date and Final Date, and it is not uncommon for the Contractor's terms to include lengthy payment periods.

There are numerous other ways in which the payment terms can be made more onerous, so do not simply accept the payment terms and timescales offered, make sure you understand them, and make sure you negotiate them!

4. The Act, and in particular the 2011 amendments to the Act, introduced some key rules about payment. Contracts that are subject to the Act are required to clearly establish two clear points in the payment process, and these are called the 'Due Date' and the 'Final Date for Payment'. It is unfortunate that in this context, the 'Due Date' does not mean the date the money is due to be paid. That is determined by the 'Final Date for Payment', which is the last date by which payment must be made.

Where the Act applies, the contract must also provide for Payment Notices to be issued by the paying party within 5 days of the Due Date, and if such notices are not provided, the Act imposes favourable default provisions. In simple terms, if your contract requires you to submit applications for payment (and most do), then your application will become the default Payment Notice. Unfortunately, the Act also introduced the concept of the Pay Less Notice!

On the plus side, this means that in the absence of a Pay Less Notice, the sum due and payable is that set out in your application. However, this favourable position is often taken away by the Contractors own terms or amendments to the Standard Form contracts.

On the down side, the dreaded Pay Less Notice is the Contractor's opportunity to make deductions from the sum he said he was going to pay you or the sum in your default notice.

5. Unless the Act applies, or the contract provides for it, there is no common law right to suspension of the works in the event of non-payment. Unless there is an express provision in the contract or the Act applies, you do not have the right to stop work if you are not getting paid. If you do stop work, you will be in breach of contract, and even if you merely threaten to stop work, it could be construed as a repudiatory breach.

The consequences of such breaches of contract can be significant, so you really do need to understand the circumstances that entitle you to suspend performance of your obligations.

Don't confuse valuation and payment. The Act basically says that if the payer does not pay what it says in the Payment Notice, then you are entitled to suspend performance. This is entirely different from the Contractor under-valuing your works.

Compare the two scenarios below:

a. Application £80,000 + Payment Notice £80,000 + Payment of £65,000 = OK to suspend

b. Application £80,000 + Payment Notice £65,000 + Payment of £65,000 = Under-valuation

Once you're certain of the terms of your contract, and circumstances allow, the right of suspension can be the most effective weapon in your fight to get paid.

6. Retention provisions should be clearly established and the triggers for payment agreed.

The deduction of retention is still common place, and despite what you might hear about campaigns to abolish it, I can't see that changing any time soon!

Retention monies are typically 3%–5% of the overall value of your contract. Usually, retention monies are released in two halves, or moieties. The first half should be released when your works reach practical completion. Some Contractors will try and introduce a delay period after your works are complete, or release retention at the Main Contract Completion Date. Don't just accept such terms, negotiate a better deal.

Ordinarily, once the Main Contract Works are complete, the first moiety of retention is released by the Employer to the Contractor. The Main Contract usually has a period from completion (usually 12 months), at the end of which the Contractor is under obligation to complete the making good of any defects. Using the JCT contract's terminology, this is the 'Defects liability period'.

When the Contractor has remedied all the defects, a 'Certificate of Making Good Defects' is issued by the Contract Administrator and the Employer releases the second moiety of retention to the Contractor. It used to be the case that the release of the second moiety of your retention was linked to the issue of such a certificate.

Where the Act applies, it precludes such arrangements and release of your retention is independent of the Main Contract. This could mean that if you complete your own defects but release of the Main Contract retention is delayed due to the failure of the Contractor to complete other defects, the Contractor would still have to pay your outstanding retention.

In order to get around these problems, Main Contractors are putting in ridiculously long timescales for release of retention. Don't just accept them, negotiate them!

Getting paid is crucial – getting angry won't help – knowing the contract will!

Valuation of variations – change of character/conditions

Most contract conditions contain provisions whereby varied work, and/or contract work affected by variations, qualifies for a revised rate if it can be demonstrated that the work has changed in character and/or conditions from that reasonably apparent from the original contract documents.

In practice, most Clients' and Contractors' quantity surveyors tend to overlook these provisions and simply use the original bill or schedule of rates rather like a 'shopping list'. This is not the correct approach. However, the initiative, in practical terms, must rest with the Subcontractor. A good question to ask is 'Did the estimator know it would be like this?' Usually, the answer is 'Not on your life'.

Leading text books emphasise that a very substantial proportion of works carried out as variations qualify for rerating. A 'checklist' summarising key factors that may be reasonably argued as having changed the original character and/or conditions of the work is given in Figure 5.1.

It can be seen from the checklist, which is by no means exhaustive, that a substantial proportion of variations in fact require rerating to reflect the changed character/conditions. Equally, it is possible to show that areas of the original contract work have been similarly affected by variations, and they may also be subject to rerating.

By maximising these avenues during the course of the works, it is possible to improve profitability and give yourself a negotiating position when seeking agreement of variation costs. This is a much better way than trying to fudge the quantities or measures, or otherwise trying to pull the wool over the QS's eyes.

However, none of these opportunities will come your way unless you take the initiative in 'flagging up' starred rates as the instructions are received and circumstances allow.

Figure 5.1 Checklist of possible factors changing character and/or conditions

1. Winter working

2. Excessive number of instructions

3. Late and/or 'piecemeal' receipt of instructions

4. Issue of instructions in disregard of agreed programmes

5. Immediate/short term response required to instructions

6. Special procurement, planning, supervision arrangements

7. Increased costs (fluctuations) on labour and materials

8. 'Piecemeal working' involving special return visits to areas in order to execute small quantities of work

9. Special isolation of individual electrical circuits in energized areas

10. Special return visits by sub-Subcontractors

11. Special hire and/or retention of plant and access equipment

12. Working in exceptionally congested circumstances due to out of sequence working

13. Working in areas that have been occupied by Employer and/or his direct contractors

14. Detours around site to gain access to obstructed areas (e.g. blocked stairways)

Don't use the bill of quantities as a 'shopping list'.

Variations – the true costs of engineering and supervision involvement

Few people on any side of the industry, least of all the Contractor's quantity surveyor, appreciate the true extent of the staff and supervision costs incurred by you, the Specialist Subcontractor, in the implementation of variation instructions. I am not referring here to the more obvious costs of installation and commissioning, but to the associated 'preliminaries' and oncosts. In fact, standard subcontract conditions often provide for such elements to be evaluated and included in the agreed valuation, if appropriate. It is up to you to bring these matters to the fore, and to 'educate' the other parties, if necessary, in order that you get properly paid for the true costs of carrying out the variation.

Figure 5.2 is a checklist of the possible range of activities that may be involved. It may open even the streetwise subbie's eyes. Each one of these activities takes time, and you are entitled to recover the cost of that time. No wonder 'variations' cost a fortune!

If there is a net addition on the eventual final account, you will of course be recovering an extra supervisory element in so far as this is deemed to be included in the rates. However, this would not normally resemble the true costs of incorporating the variations, which often include many omission items. Indeed, when you make a financial 'omission', you are actually deducting a sum of money that usually includes a supervision element in the rates. So, you are being hit with a 'double whammy', since the 'omission' still has to be processed and evaluated.

When we submitted a case to an adjudicator based on this approach, the result was a resounding success. You will be glad to know that the financial results made our exercise all worthwhile, not least because it added over £30,000 to the amount the Specialist Subcontractor received.

So, there you have it. It really is up to you to take the lead, as early as possible in the job. Hopefully, my little 'checklist' will help, and if you would like me to send you a spreadsheet incorporating the above, plus a series of columns to form a matrix for valuation, please get in touch.

Figure 5.2 Variations – a checklist of engineering and supervisory involvement.

1. Receive and process revised/additional drawing including issue to site and to any affected specialists/suppliers/manufactures.

2. Examine drawings for changes, omissions and/or additions.

3. Encounter change/omission/addition/problem/deficiency on drawings and/or other design details.

4. Encounter ditto at work face.

5. If necessary, find alternative work for operatives, including all necessary liaison and arrangements with Contractor regarding access/scaffolding/coordination with other trades etc. in new location/s.

6. Instruct operatives regarding alternative work and supervise relocation.

7. Make out and submit technical query sheet.

8. Record and submit confirmation of instruction.

9. Telephone discussion/arrange meeting on-site with Contractor/consulting engineer/clerk of works etc.

10. Attend site discussion as Activity 9.

11. Confirm outcome of discussion in writing.

12. Measure at work face, prepare and submit detailed sketches for installation (e.g. hangers and brackets in ceilings).

13. Obtain formal written direction and process.

14. Mark up drawings and mark out at work face.

15. Cancel and/or amend orders for materials, equipment and/or plant.

16. Arrange return/transport of original materials, equipment, and/or plant to stores and/or to specialist supplier/manufacturer.

17. Arrange and attend all necessary meetings with specialist suppliers and/or manufacturers to discuss implications and/or product solutions.

18. Obtain quotation/s for revised materials/equipment. Negotiate reduction if possible.

19. Obtain approval of Contractor/quantity surveyor/consulting engineer to said quotations.

20. Place order/s with suppliers and/or manufacturers.

21. Rearrange programmes, both short and long term.
22. Notify Contractor of delays/disturbances/costs by site delay notice/email/letter.
23. Update Activity22 weekly.
24. Obtain formal written directions, process the same and record within site and office filing system.
25. Make special arrangements with Contractor for access/attendance/scaffolding/removal of same/coordination with other trades, in connection with Activities 17–24 inclusive.
26. Where necessary, arrange to disconnect power from areas to be worked in, and reconnect after completion, including discussing with Contractor any effect on other trades.
27. Organise, obtain/relocate labour and plant resources in return to original affected work face.
28. Instruct and supervise labour upon return to original affected work face.
29. Arrange, supervise the retest and commissioning of affected circuits where necessary, after execution of varied works.
30. Modify as fitted drawings to incorporate variations/additions/omissions.
31. Note abortive engineering and supervision time already spent on any omitted works.
32. Receive, check and process supplier/manufacturer's invoice for varied works.
33. Site measure for progress/productivity records and for interim application.
34. Measure revised quantities from drawings.
35. Evaluate revised quantities by reference to bill or 'fair valuation'.
36. Submit revised quantities/valuation for approval by Contractor/quantity surveyor.
37. Include revised valuation in interim application.
38. Incorporate in management accounts system.
39. Pursue interim agreement/payment via telephone/correspondence/meetings.
40. Include revised quantities in final account and pursue agreement/payment.

Take the initiative in explaining and claiming the true costs of dealing with variations.

Are your systems user-friendly?

In recent years, even the most modest office has undergone a revolution in terms of 'systems'. I mean things such as wages records, invoices, purchases, plant records and orders. Computers and software have become very affordable, and there is no shortage of people with the necessary computer skills. Most of this reorganisation of the office has been entrusted to bright young accountant types. One result has been that the systems are entirely 'accountancy driven' (i.e. devised by and for the convenience of accountants). On the surface this does not appear to present a problem.

It is not until some buffoon such as myself asks or looks for information in order to pursue reimbursement of variations, claims, etc. that the shortcomings become painfully apparent. Let me give some examples. Total worked hours on the project? System doesn't tell you, but only paid hours – utterly irrelevant. Cost of supervisory staff? Sorry – included in with operatives. Plant hire invoices for the project? Sorry, would need to be extracted from the general heap for all jobs. Cumulative expenditure on project to date for individual suppliers? Sorry, each month's materials, although detailed, are added back onto the previous cumulative figure for all materials. And so it goes…

The result is to add greatly to the time and cost involved in pursuing claims for variations and delays. This is particularly true in the case of the 'further and better particulars' so beloved of the pink-cheeked young quantity surveyors employed by today's Contractors. 'Chasing payment? I'm still waiting for the information to check your claim.' Very often, the job is given up as hopeless or monies go outstanding for years.

So why not make your systems 'objective friendly'? In other words, gear the systems to the end objective, which is to get paid for what you do?

> **Systems should help to recover your entitlements – not please the system designer!**

*S*ection 6

Minimise confrontation

Claims and confrontation

Perhaps no word arouses such an immediate emotional reaction in our industry as that of 'claims'. How silly and illogical. Let me just put the subject in perspective. Firstly, many of the 'relevant events' that are now listed in the standard conditions as qualifying for extension of time and/or financial reimbursement are, in essence, 'breaches of contract'. I am referring to such things as failure to give possession, denial of access and delay in provision of necessary information. However, over the years, as such breaches became increasingly commonplace, the standard conditions have been amended to 'legitimise' these breaches and to provide remedies and entitlements under the contract for the injured party. That is why the standard forms now list such a multitude of these events.

So, what we are looking at in the standard forms is a list of 'breaches' which the Client/Contractor is now licensed, not to say encouraged, to commit at will, and a list of prescribed remedies for the guy on the receiving end (in our case you!). These principal remedies take the form of extension of the completion date (thereby protecting you from damages) and the reimbursement of loss and expense. They also have the effect of maintaining a definite completion date for the purposes of liquidated damages deductions by the Client, but this is another subject. These standard forms lay down not only the remedies, but a set of detailed procedures which the injured party must follow in order to 'claim' his entitlements. These principally concern giving timely written notice and submission of detailed records.

So why should there be any question of 'confrontation'? The injured party is merely following the procedures which have been forced upon him. Would Manchester United be accused of 'confrontation' if they demanded 'compensation' in the event that Sir Alec Ferguson had been enticed away by another club during his reign at old Trafford?

Most Specialist Subcontractors would prefer to be given access and information on time, do the job on time, get paid and walk happily away. It is not, in most cases, their fault that they are caught up in delays and disruption due to events that once would have been termed 'breaches of contract'.

It may therefore be seen that those Clients and Contractors who choose to complain of 'confrontation' have never taken the trouble to think things through in an adult manner. Indeed, their cries of 'confrontation' are in many cases a defensive strategy, by which it is hoped to frighten the injured subbie from following the set procedures for 'compensation'. If this approach is to be followed, then should not the subbie cry 'confrontation' every time he is denied access or information, and/or receives yet another late variation? If he did so, his voice would soon become hoarse!

So, yet again, my advice is to stay calm, co-operative and professional in all your dealings, including that of recovering your just entitlements. I have proved, over the years, that by following this 'adult' approach, any confrontation can be kept to a minimum.

Above all, remember that prompt notices and good records are essential if you are to secure your just entitlements.

Stay calm, co-operative and professional to minimise any confrontation.

Damages and extension of time

One of the biggest financial risks to a Specialist Subcontractor is the possibility of damages from the Client and/or Contractor if you overrun your original completion date. In most cases, the end Client will be automatically entitled to a fixed weekly rate of liquidated damages from the Contractor, who will then seek to pass these down to all the other Specialist Subcontractors, along with the Contractor's own prolongation costs.

These overrun costs can be massive. Consider a contract with liquidated damages from the Client of £10,000 per week, plus the weekly costs of the Contractor's site staff and establishment, and also those of other Subcontractors. You could be looking at £30,000 per week of overrun being deducted from your account and/or pursued from you through adjudication or litigation. Enough to sink many a smaller subbie for good.

If you wait until disaster strikes before considering what you should do, then you are in big trouble. It's probably too late by then in most amended subcontracts, and definitely those incorporating NEC/4 (previously NEC/3)!

In JCT contracts, what you must do is to give the Contractor written notice of individual delays immediately they become apparent, detailing the cause and the effect upon your programme (in NEC/4 contracts the situation is altogether different, so please check out Section 8). In particular, you must notify the anticipated effect upon your completion date and request an extension of time. Then you must update these notices as the works proceed.

As and when you consider the original completion date to have become unachievable, you should request an extension of your original period. The matters giving rise to entitlement to extension of time are usually listed in the subcontract documents, and in JCT contracts are termed 'relevant events'. In certain cases, prolongation caused by such an event is reimbursable as a 'relevant matter'. In other cases, the event is 'neutral' and therefore there may be time but not money.

The following is a list of the main relevant events usually to be found in a standard form of building contract or subcontract. Those which are eligible for financial compensation are shown as emboldened. However, be aware that many 'bespoke' forms produced by Clients and Contractors are far less generous and the list will be much shorter!

- **Architect's instructions or variations**
- **Delayed receipt of necessary instructions or information** (*but only if you have requested same in writing and not too early and not too late in relation to the programme/progress of your works*)
- **Delay by the Employer/Client or his supplier in supply of goods or materials**
- **Delay by the Employer/Client or his direct Contractor in execution of works**
- **Execution of works by the Employer/Client or his direct Contractor**
- Delays due to statutory authorities
- Strikes and civil commotion
- Exceptionally adverse weather
- Force majeure
- Terrorism or threat of terrorism

A further source of entitlement arises from default on the part of the Contractor and/or his other Subcontractors (i.e. delayed availability of workfaces, lack of building progress and weatherproofing, etc.). These types of preventions are usually classed as '**act, omission or default**'. Such matters are eligible for financial compensation. However, it is very difficult to get money out of the Contractor's own pocket, and it is wise to run this claim parallel with those which point towards the end Client, in which case the Contractor will be more receptive.

You must also keep and regularly submit good site records (i.e. site diary, progress reports, etc.) in order to prove your entitlements when challenged. There is no need for all these notices and records to read like some legalistic document. Just tell it like it is, making reference to the facts and the anticipated effects.

But you must submit these details in writing and as soon as you encounter the problem. You must also ensure that you comply with the specific requirements of the contract. And these can be onerous. If it says you must send such notices on blue paper in a green envelope tied up with pink ribbon, then that is what you must do to ensure the notice is valid!

If you carry out these obligations, and providing you are using your best endeavours (i.e. doing your best) on-site, then you are entitled to an extension of time, thus protecting you from damages. But don't be surprised if the Contractor doesn't actually grant you an extension of time. They rarely do, but what you are doing is ensuring that you amass sufficient evidence to secure your entitlement formally should that ever be necessary.

If you have the notices, records and evidence in place, then even if the Contractor comes at you, it should be possible to win in the end. So, to quote a well-worn phrase, 'Don't be shy – get noticed'.

Remember, no delay notice means no extension of time!

Claim for delay costs

Most subcontract conditions provide for payment of loss and expense (i.e. 'prolongation costs') in the event of the Specialist Subcontractor being delayed beyond your agreed original completion date.

However, any entitlement is strictly dependent upon the following:

- The reasons for the delay must be such as to give rise to a legitimate entitlement under the terms of the subcontract agreement or order.
- You must make written application for loss and expense as soon as the likelihood of incurring costs becomes apparent.
- You must be able to provide evidence as to the circumstances.
- You must demonstrate the 'cause and effect' (i.e. the clear link between each notified event, the delay and the resultant loss and expense).
- You must state the contractual basis of his claim (i.e. which clauses are relied upon, and/or common law).
- You must provide details and substantiate your actual loss and expense.

If these criteria are satisfied, there should be no need for 'confrontation'. The key circumstances giving rise to a delay claim usually include late issue of information, late and/or excessive variations, discrepancies in the documents, obstructions by the Client/Employer and/or his direct Contractors and default by the Contractor or other subbies. The latter would include late access, delays by preceding trades, etc.

As soon as you becomes aware of any such event, you must make written application specifying the circumstances, the effect upon programme and completion, and the fact or likelihood of loss and expense (i.e. additional costs). Fuller details of these financial consequences should then be submitted as quickly as possible. Furthermore, a sum should be included in the very next interim application after the costs begin to accrue. If you think you do not need to do this immediately, then check your contract because it is likely that immediate notice will be a 'condition precedent' to your right to recover those costs. In other words, 'no notice – no recovery'.

The costs of delayed completion will usually be expressed as **weekly time-related actual costs** such as supervisors, chargehands, stores and welfare attendants, site cabins and facilities, scaffold, plant and equipment (often jointly referred to as 'preliminaries'). Also, claimable will be the prolonged involvement of engineers, quantity surveyors and contract managers. These costs are usually detailed on the basis of a weekly total and then multiplied by the number of weeks overrun. It may be that the installation period has been prolonged by a greater period than that of the final overrun (e.g. the 'commissioning only' period has disappeared). In this case, providing this is due to eligible events, you should claim accordingly.

Despite the common view held by many quantity surveyors, the loss and expense of subcontract overrun should be that of 'actual cost' and not by reference to the bill rates. A checklist of typical prolongation costs for a medium-sized project is appended below. The rates used are purely for example.

It is also customary to claim for **head office overheads**. One approach is to add these as a percentage to the net costs of the over-run. An alternative is that of the 'Hudson formula' or 'Emden formula', but these are unlikely to find favour. A successful claim for such costs requires hard evidence of company accounts, annual turnover, head office staffing levels and the like (see 'Head office headaches').

On fixed price jobs with substantial overruns, labour and material costs may have escalated due to a wage increase and/or the volatility of material prices such as that of copper. In these cases, it is usually necessary to demonstrate specific increases by means of national working wage agreements, wage sheets, purchase orders and invoices.

Details	Name	Weekly involvement (%)	Weekly rate	Cost per week
Personnel				
Site engineer	F. Frazer	100	1075.00	1075.00
Site supervisor	T. Jones	100	765.00	765.00
Chargehand	A. Bloggs	100	685.00	685.00
Chargehand	T. Clancy	50	660.00	330.00
Welfare/stores man	F. Lee	50	460.00	230.00
Quantity surveyor	B. Sharpe	50	1020.00	510.00
Contracts manager	I. Teflon	40	1150.00	575.00
Total				**£4170.00**
Accommodation				
Site office	1 No		65.00	65.00
Mess cabin	1 No		45.00	45.00
Storage containers	2 No		35.00	70.00

(Continued)

(*Continued*)

Details	Name	Weekly involvement (%)	Weekly rate	Cost per week
Telephone and calls	1 No		4500	45.00
Photocopier	1 No		10.00	10.00
Site computer	1 No		25.00	25.00
Mobile phones and calls	1 No		25.00	25.00
Office furniture	Lot		40.00	40.00
Total				**£325.00**
Plant and tools				
Site van and fuel	1 No		120.00	120.00
MEWP	2 No		175.00	350.00
Tower scaffold	2 No		25.00	50.00
Drills and leads	12 No		16.00	192.00
Testing equipment	Lot		50.00	50.00
Total				**£762.00**
Collection				
Personnel				4170.00
Accommodation				325.00
Plant and tools				762.00
Weekly total cost				**£5257.00**

A further claimable element is that of **finance costs**. These are usually allowed on the basis of an appropriate percentage above base rate (to reflect your financing arrangements), compounded quarterly, this rate being applied to the total of the loss and expense, from the midpoint of the overrun period up to the date of the claim. This claim for finance costs should then be updated on a monthly basis as long as the claim goes unresolved.

You must give early written notice of your claim and make formal application, with a budget figure pending detailed submissions. This will not make you very popular, but it will prevent you from losing your entitlements due to lack of notice.

> **Identify, notify, apply for and evaluate delay claims as they become apparent.**

Head office headaches

If your original contract period is prolonged by certain qualifying events (e.g. variations, late information and delayed access), then you are entitled to be paid for the additional costs of your time-related staff, plant and accommodation. However, what about the often-thorny subject of 'head office overheads'? For some reason, this claim heading is habitually difficult to prove, and you recover less than your proper entitlements. Why should this be?

Firstly, do we ourselves believe in our case? Well, there is no doubt whatever that any substantial overrun of a decent sized job will similarly prolong the involvement of office-based personnel (i.e. contracts managers, engineers, buyers, wages and accounts clerks).

One approach to recovery is by way of a formula ('Hudson' or 'Emden'). This certainly produces the biggest numbers, but their crude application and lack of proof usually make their use a 'non-starter'. You are far more likely to secure agreement with a claim for specific additional involvement of managers and office staff.

And, proving your rights need not be hard. The biggest single step is to set up a system whereby all staff, from directors downwards, allocate their time to individual subcontracts on a daily basis. This time should then be costed to the appropriate job. Once you have established the additional costs of your time-related head office staff, it would then be reasonable to add a reduced percentage to your total claimed costs to cover for your fixed overheads, such as lease of premises, rates and gas and electricity charges.

The approach recommended above may not produce the 'monopoly' numbers which accumulate when you use a formula approach, but people will take you more seriously and you stand a much better chance of getting some recovery for your head office overheads.

> **Directors and staff should allocate their time to individual projects.**

Disruption claims

Disruption can occur whether or not there is any delay to progress. In simplest terms disruption could be described as 'being messed about'. The result is usually a reduction in planned labour productivity. The scenario includes one or more of the following: interrupted and out of sequence working, piecemeal activity, return visits, excessive overtime working and congested workfaces. Many subbies have been driven to insolvency this way!

The typical causes include inadequate building progress, lack of progress by other trades, late information, late and/or excessive variations, delays or obstructions by the Client or his direct Contractors. Another cause can be compressed working enforced by the Contractor in an endeavour to catch up lost progress. These and other circumstances are listed in most standard forms of subcontract as matters giving rise to an entitlement to reimbursement of the loss and/or expense incurred.

However, there are some big hurdles to be cleared. First, you must give written notice as soon as you foresee the problem and promptly follow up with a formal application for reimbursement of loss and expense. The notice and application must detail the circumstances, the location, the likely effect on your programme and the estimated effect upon completion. In theory at least, this is so that the Contractor can address the problem and minimise or avoid the same. But, more particularly it is to overcome the 'no notice, no entitlement' clauses you will find in most Contractor's contracts.

Regrettably, it is more common for the subbie's warnings to be rejected, sometimes with accompanying hostility. You must stay cool and be professional. The real difficulty is that 'he who alleges must prove', therefore you have to prove your case and evaluate the additional costs in a manner that will command acceptance. There has been much discussion and case law surrounding what are termed 'global claims' (i.e. roll up your costs and claim the overall loss). Whether we like it or not, we now have to be prepared to set out detailed **'cause and effect'**. This means being able to show how each principal event affected the programme and the precise costs that flowed from that event. In the simplest of cases, this involves daywork records of men standing or relocating. However, it is more often a case of reduced productivity in actual working time.

You have little real chance of demonstrating your entitlement, unless you have kept **detailed labour allocations, dates and durations of programme interruptions,**

progress records and a good site diary, all as a matter of course. The history of individual programme activities can then be plotted on a simple bar chart, comparing planned with actual progress and labour resources, together with coded cross references to the various problems (i.e. architect's instructions, receipt of key drawings, denial of access, etc.).

Another approach is to demonstrate the productivity achieved on a 'good' but 'typical' area, and to compare this with the productivity achieved on the 'bad' areas (i.e. where 'good' means relatively undisrupted and 'bad' means 'disrupted'). Such an exercise can provide powerful evidence of your ability to perform to your planned norms, when given the chance, and when coupled with strong evidence of the disruption, this can form a sound basis for financial evaluation. It is also an approach which has received judicial approval in the past.

Other evaluations can be based upon researched percentages of productivity loss resulting from such matters as **extended overtime working, increased gang strengths and congested workfaces**. There is an excellent book published by the Chartered Institute of Building, entitled *Effects of Accelerated Working, Delays and Disruption on Labour Productivity*. Indeed, some of the conclusions are mind-boggling. Having submitted your claim, hang on in there. Nobody said it would be easy!

Identify and notify disruption and keep good records.

Key points of the 'Delay and Disruption Protocol' – ignore it at your peril!

The Society of Construction Law's (SCL) Delay and Disruption Protocol was first published in 2002. The intention was to provide a scheme whereby delay could be better controlled and managed during the construction process. Following the publication on the 1 July 2015 of an intermediary update known as Rider 1, the Second Edition of the Protocol was release in February 2017.

In updating the Protocol, there were eight specific terms of reference:

1. Whether the expressed preference should remain for time-impact analysis as a programming methodology where the effects of delay events are known

2. The menu and descriptions of delay methodologies for after the event analysis

3. Whether the Protocol should identify case law (the United Kingdom and international) that has referenced the Protocol

4. Record keeping

5. Global claims and concurrent delay

6. Approach to consideration of claims (prolongation/disruption – time and money) during currency of project

7. Model clauses

8. Disruption

Legal status of the Protocol

Unless it is incorporated in the contract, the Protocol has no force of law. However, it has been used as a benchmark for how to approach delay analysis, and I regularly refer to it. Especially when Contractors have made what amounts to a crude analysis of the delay that they are blaming on one of my Subcontractor Clients!

Forms of delay analysis

The Second Edition refers to prompt, and contemporary, evaluation. The new Core Principle 4 notes as follows: *"Do not 'wait and see' regarding impact of delay events (contemporaneous analysis) The parties should attempt so far as possible to deal with the time impact of Employer Risk Events as the work proceeds (both in terms of EOT and compensation). Applications for an EOT should be made and dealt with as close in time as possible to the delay event that gives rise to the application. A 'wait and see' approach to assessing EOT is discouraged".*

The original Protocol recommended that one particular form of delay analysis, namely the time-impact form of delay analysis methodology, be used wherever the circumstances permitted. One of the main reasons for the review of the Protocol was that this was not universally supported. Whereas the original Protocol made no mention of the 'windows' form of delay analysis, it has become one of the most used forms of delay analysis, arguably because it is considered to be one of the most reliable.

In the Second Edition of the Protocol, no one form of delay analysis is preferred, instead, the Second Edition sets out the factors that need to be taken into account in selecting the most appropriate form of delay analysis as well as providing a helpful explanation of many of the delay analysis methodologies currently in common use.

Concurrent delay

The SCL say that the approach to concurrent delay in the original Protocol has been amended in the Second Edition to reflect recent case law. The Second Edition defines concurrent delay in this way:

> True concurrent delay is the occurrence of two or more delay events at the same time, one an Employer Risk Event, the other a Contractor Risk Event, and the effects of which are felt at the same time. For concurrent delay to exist, each of the Employer Risk Event and the Contractor Risk Event must be an effective cause of Delay to Completion (i.e. the delays must both affect the critical path). Where Contractor Delay to Completion occurs or has an effect concurrently with Employer Delay to Completion, the Contractor's concurrent delay should not reduce any EOT due.

The Second Edition therefore recognises that true concurrency is rare, and also notes that where Employer Risk Events and Contractor Risk Events occur sequentially but have concurrent effects, the delay analysis should determine whether there is concurrent delay and, if so, whether an extension of time is due for the period of that concurrency.

The approach to notices

Elsewhere in this book I have stressed the importance of complying with contractual notice procedures. This is, unsurprisingly, endorsed by Rider 1 which stresses that

> The parties and the CA should comply with the contractual procedural requirements relating to notices, particulars, substantiation and assessment in relation to delay events...

Global claims

The Second Edition has this to say about global claims:

> The not uncommon practice of contractors making composite or global claims without attempting to substantiate cause and effect is discouraged by the Protocol, despite an apparent trend for the courts to take a more lenient approach when considering global claims.

It goes on to say that Contractors should be aware that there is a risk that a global claim will fail entirely, if any material part of the global loss can be shown to have been caused by a factor or factors for which the Employer bears no responsibility. The Contractor must try to provide adequate records to enable the Engineer or other adjudicator to establish a causal link between the Employer's Risk Event and any resultant costs or delay.

Records

This further confirms the importance of maintaining records. Appendix B of the Protocol lists record types relevant to delay and disruption. Further, Core Principle 1 of the Second Edition notes that

> Contracting parties should reach a clear agreement on the type of records to be kept and allocate the necessary resources to meet that agreement.

The Protocol provides guidelines on the keeping of records and advises that in order to avoid disputes, where practicable, records should be signed by representatives of the Employer and Contractor. The Protocol recognises that there is a cost here and specifically notes that

> Good record keeping requires an investment of time and cost, and the commitment of staff resources by all project participants. It is therefore recommended that, prior to preparing the tender documents, the Employer considers its requirements of the Contractor in relation to record keeping and includes these within the tender documents.

Whether the benefit of having better records means that the Protocol and detailed record keeping requirements will find their way into contracts as standard remains to be seen.

Conclusion

One cannot quarrel with the commendable aims of the Protocol that continues to have value and an increasing influence by providing guidance as to good (and even best) practice. Its fundamental starting point, namely 'that transparency of information and methodology is central to both dispute prevention and dispute resolution', is universal.

The emphasis on programming and early identification and management of problems is very similar to the increasingly used NEC contract. If a Subcontractor diligently follows the recommendations from the outset of a project, he will be in a much stronger position to protect his entitlements.

However, if the Protocol recommendations are to be diligently followed, there will be a need for additional staff on site to handle the demanding regime of notices, records and forecasts. In particular, there is a need for more planners. The question is whether most Subcontractors can accommodate the extra costs required to operate the procedures. On the other hand, if the procedures are ignored, then you may well lose your entitlements.

Gear up from day one to comply with the Protocol!

Loss and expense – some questions answered

Here are answers to some frequent questions which arise regarding loss and expense. But be careful, because the answers are based on the Standard Form subcontracts, e.g. JCT. If you have signed up to onerous terms, then your entitlements can be significantly different.

Can a Contractor refuse to value your loss and expense because the architect has refused to certify loss and expense under the Main Contract?

Under standard forms such as JCT, this approach is invalid. The Contractor may be in default on the Main Contract, with no entitlement. You may well be genuinely entitled to a full recovery under the subcontract. But, beware of bespoke conditions and amendments to standard forms limiting your entitlements to whatever the Contractor can recover under the Main Contract.

What if the main reason for delay is lack of building progress and/or access?

Under most standard forms, you would have entitlements due to the
'act, omission or default' of the Contractor. Again, beware contracts that omit such provisions.

Clients and Contractors usually want to base prolongation costs (i.e. job overrun) on the allowances made in the Subcontractor's tender. Is this correct?

Unless there are onerous provisions to the contrary, the correct approach is to value site establishment, staff and plant on the basis of actual costs properly incurred as a direct result of the delay.

At what point in the programme should costs of prolongation be valued?

The assessment should be made at the point where the delay took place. It is more accurate than valuing the 'tailend' overrun. One very good reason for this is that the tail-end resources have often been reduced, but this would always have been the case. The delays usually occur throughout the course of the job. This may involve a series of calculations for varying periods of delay at differing points in the progress of the works.

What is meant by 'prelims thickening'?

This term refers to increase in weekly involvement of management, technical and supervisory resources, caused by volume of variations or increased difficulties imposed upon the programme. This often happens during a period of intense disruption to the programme. If you have to retain or introduce additional engineers, charge-hands, etc., you should advise the Contractor at the time, explaining the reasons and stating your intention to seek financial compensation. One effective way of valuing 'thickening' is by a histogram showing weekly details of planned and actual 'prelims' resources, supported by a daily site register showing all persons, including office staff.

What is the best way of claiming for 'off-site' overheads and staff due to prolongation?

One way is the use of a formula approach (e.g. Hudson or Emden), but it is unlikely to succeed on its own. A better approach is to ensure that all off-site managers and staff allocate their time to individual projects. This time can then be proved as a genuine cost. This method has the added benefit of reducing the residual percentage for fixed overheads (e.g. office lease, rates, heat and light). This latter can then be added, usually as a percentage.

What is meant by 'time at large'?

Time may arguably be set at large when an act of prevention by the Client or Contractor causes delay to completion, and the contract contains no specific mechanism for revision of the completion date. A typical example occurs in bespoke contracts, those used by Clients for purchase of goods or services, or where the Contractor deliberately omits provision for extension due to his own 'act, omission or default'. Your obligation would then be to complete in a 'reasonable time'.

What is the best way to safeguard entitlements to time and money?

The best safeguard is to give prompt written notice of all delays and make prompt written application in respect of all additional costs as soon as the likelihood becomes apparent, and to maintain good records. Keep the Client or Contractor informed at all times and maintain a proactive attitude. All too many subcontractors 'leave it until the end' before claiming their additional costs, for fear of upsetting the Contractor. This is foolhardy in the extreme, and they can hardly complain when their claims are rejected.

Ensure you know how to recover loss and/or expense

Set-off and contras

In the Third Edition, Jack said, '*Some builders have a policy of recouping their losses by setting off money from their subbies. Out of the blue comes a letter from the builder blaming our unfortunate subbie for the project delays, and informing him that the costs of liquidated damages and the builder's own prelims costs will be deducted (i.e. "set off") from the next payment due*'.

Sadly, that situation has become ever more prevalent, and it seems that the only way that some Contractors can return a profit is to beat it out of their Subcontractors. Of course, not all Contractors operate in this way, but let's just say that there is not a week goes by when we are not advising our exclusively Subcontractor Clients about such unfair treatment!

A good time to tackle these problems is during the 'tender to order process' when you have the opportunity to review and negotiate the subcontract conditions.

Many set-off clauses (i.e. as found on the Contractor's own onerous forms) are a 'license to print money'. Most major Contractors have clauses entitling them to deduct set-off for future estimated costs that may or may not actually happen. Other fertile opportunities for contras may be based on alleged damage to other trades, additional attendance provided by the Contractor, failure to clear rubbish, etc. (e.g. 'footprints on the carpet' and 'fingerprints on the ceiling'). Many also claim the right to deduct money in respect of other subcontracts.

If you are prepared to sign up on terms like that, you are asking for trouble and will probably get it.

The Construction Act requires that prior written notice must be given by any party who intends to withhold money from payments otherwise due, i.e. the dreaded Pay Less Notice. The most common standard forms JCT SBCSub/C and DBSub/C at Clause 4.10.5 require the Contractor to serve any such notice not later than 5 days before the 'Final Date' for the next interim payment. The first thing most Contractors do is reduce that period to 1 day. And, whilst that notice requires the Contractor to specify 'the basis on which that sum has been calculated', it regrettably does not require much effort to meet that 'test'.

Despite the length of time the Construction Act has been around, there are still some Contractors who are oblivious to its requirements, whether belligerently or ignorantly. So, if a Contractor deducts contras without notice, one option is to threaten adjudication (assuming the Act applies or there are express terms). You should take professional advice first, but in a blatant case of abuse of the Act, one would hope for a successful and speedy outcome.

During the job itself, make sure you operate good 'housekeeping' (i.e. prompt removal of your trade waste from working areas and care of your plant and equipment). Ensure that you respond quickly to the Contractor's complaints regarding site cleaning or damage. If you are responsible, then attend to the matter forthwith. Take photographs of completed areas, and keep records to protect yourself. If possible, try to get the Contractor's signature for handovers of such areas.

Above all, you must deal promptly with any delay to your works and submit delay notices in writing at the time, and request an extension of time if necessary. Please see Section 2.

Many Contractors wait until the end of the job before producing their claims against you, supported by a bundle of ancient daywork sheets going back several months. This is a common negotiating tactic aimed simply and crudely at reducing your account. Do not stand for it, it is probably a 'try on'. Take action immediately and if necessary escalate the matter, and if the sum of money is big enough, take professional advice. In a good many cases, our involvement and the veiled threat of adjudication is enough to cause a bullying Contractor to 'back off'.

Stand up for your rights – don't just accept what the Contractor says!

Section 7

The Construction Act

Know your rights

Introduction to the Construction Act

The Housing Grants, Construction and Regeneration Act 1996 came into force on all new construction contracts entered into after 1 May 1998. The Act was amended by The Local Democracy, Economic Development and Construction Act 2009 and covers both Scotland and England, it amends the Housing Grants, Construction and Redevelopment Act 1996. It came into force in England on 1 October 2011 and on 1 November 2011 in Scotland.

The Act embraces all normal construction operations including labour-only contracts, site clearance, demolition, repair works, landscaping and consultancy agreements such as architects and consulting engineers. However, some activities are not 'construction operations' as defined in the Act and are therefore excluded. These are supply-only contracts, supply and fixing of plant in process industries, contracts with residential occupiers, PFI contracts (but not the construction contracts entered into pursuant to them) and certain other definitions.

Sweeping reforms were introduced, on the back of the Latham report 'Constructing The Team', aimed at removing the worst of the injustices, which caused so much trouble in the construction industry. These reforms fall under two prime headings, namely '**payment**' and '**adjudication**'.

The Act and the accompanying 'Scheme for Construction Contracts' have each been the subject of criticism, and although they were amended in 2011, the amendments didn't fix all the problems. However, since the introduction of a statutory right to adjudication, literally tens if not hundreds of thousands of cases have been decided in that way, with the RICS alone making thousands of adjudicator appointments every year.

As to payment, the minimum requirements of the Act are rather modest, albeit that the 2011 amendments did make certain improvements. In some ways, the Act offers less protection than that which was formerly available in the industry's standard forms of contract. It also the case that there are many 'loopholes' that enable a party to frame their conditions of contract in order to evade the true spirit of the Act. Later in this section, we summarise some of the 'dirty tricks and dodges' being used by various Contractors.

In the Third Edition, Jack suggested that *'the Act represents, along with the previous Latham initiatives, a continuing "wind of change" that is blowing throughout the industry'* and that *'One way or the other, it is going to become progressively harder to operate like Al Capone and those who do so will probably incur a lot of unfavourable publicity in the trade press'*.

Were he alive today, I'm sure Jack would be dismayed to see that his optimism was ill founded, and that 'Subby Bashing' is alive and well. And as for the Government and the Trade Associations efforts, a major survey of some 355 Specialist Contractors, supported by Streetwisesubbie.com, in 2017 found that in the opinion of 93.23% of the respondents, Trade Associations and Government were NOT doing enough to resolve payment problems.

As a professional Consultant, the author is continually advising Specialist Contractors regarding their entitlements under the Construction Act. Whilst the following pages seek to highlight the main features in a practical way, we strongly recommend taking professional advice if you have a problem. It is all too easy to make matters worse if you do not!

Know your rights under the Construction Act, and use them!

Payment under the Construction Act

The basic payment rules introduced by the Construction Act and the 2009 amendments are

1. The right to payment by instalments.

2. An 'adequate mechanism' for determining what sums are due and when, and linking to the 'performance of obligations' or 'decisions' under 'another contract' does not constitute an 'adequate mechanism'.

3. Prior notice of sums due and 'the basis on which that sum is calculated'.

4. 'Pay when paid' clauses to be 'ineffective' (except in the case of insolvency of a third party upon whom payment depends).

5. Prior notice of intention to pay less than (i.e. 'set off from') the notified sum setting out 'the basis on which that sum is calculated'.

6. Right to suspend work (by not less than 7 days' notice) for non-payment of the 'notified' sum.

Where one or more of these minimum requirements are not met, and/or no agreement has been reached on the terms, the relevant parts of the Scheme for Construction Contracts come into operation as a 'default' mechanism.

The Act leaves the parties 'free to agree' the amount of any instalments or periodic payments, the mechanism for determining this, the intervals at which such payments become due and the intervals between the 'Due Date' and the 'Final Date' (i.e. the latest date by which payment must be made). This allows Contractors to use their 'muscle' in order to impose longer payment periods than are fair and reasonable.

The Local Democracy Economic Development and Construction Act 2009 ('LDEDC') introduced changes to the law regarding payments under construction contracts for contracts entered into on or after 1 October 2011.

Stage payments

The Act (see section 109 HGCR Act) entitles the payee to stage payments for any work under the contract unless

- The contract states that the duration of the work is to be less than 45 days, or
- The parties agree that the duration of the work is estimated to be less than 45 days.

Due Dates and Final Dates

Every construction contract must provide an adequate mechanism for determining what payments become due under the contract, and when, and provide a final date for payment of any sum that becomes due. The parties are free to agree the due dates for payment, and how long the period is to be between the date a sum becomes due and the final date for payment (see s110).

Payments cannot be made conditional on the payer receiving payment from a third party (see s113), unless the third party paying the payer is insolvent (the Act defines what constitutes insolvency for the purposes of this provision).

The LDEDC has also prohibited payments being made conditional on the performance of obligations under another contract or a decision by any person as to whether obligations under another contract have been performed. Therefore, a Contractor cannot make payments to his Subcontractors conditional on the Employer certifying his own payments as being due under the Main Contract.

What happens if my contract does not comply with the Act?

If the contract does not provide an adequate mechanism for determining when payments become due and the final date for payment, the 'Scheme for Construction Contracts' applies. The Scheme is discussed further below.

Who notifies the amount due for payment, and how?

The LDEDC has introduced a new payment regime. Previously, only the payer could serve a notice stating how much he was due. Under the new provisions, the contract can specify who gives the notice of what is due. It can be

1. The payer, or

2. Another 'specified person', this might be a Project Manager or Contract Administrator, for example or

3. It could be the payee who gives the notice.

The notice must be given not later than 5 days after the payment due date. In each case, it must state the sum the person giving the notice considers to be due or to have been due at the payment due date and the basis on which the sum is calculated (see s110A). It is 'immaterial' that the sum due is zero, a notice must still be given.

If the payer or the specified person is supposed to serve the notice under the contract and he does not, then the Act provides that the payee can serve a notice instead, stating the amount he considers due and the basis on which that is calculated (see s110B).

Default payment notice

Importantly, if the contract permits or requires the payee (prior to the date on which the payee's notice is required to be given) to notify the payer or a specified person of the sum that the payee considers will become due on the payment due date and the basis on which that sum is calculated, and the payee gives such notification, e.g. in the form of a payment application, then this is to be regarded as the payee's notice and becomes what has come to be known as the 'default payment notice' and the payee cannot give another notice.

Contract does not provide for 'payee's notice' (interim application)

If there is no provision in the contract for a payee notice (interim application), then the payee can serve a notice under section 110B (2), and the final date for payment is pushed back by the number of days between the notice that should have been given and the payee's notice.

The amount in the notice given (whether by the payer, payee or specified person) must be paid on or before the final date for payment, unless a notice to pay less has been served by the payer.

Payless notice

In order to withhold monies from the payee, the payer (or the specified person) must serve a 'pay less' notice (this used to be known as a 'withholding notice'), see s111. The pay less notice must

- Specify the sum the payer considers to be due on the date the notice is served.
- The basis on which that sum is calculated.

This is slightly different to the old withholding notices and has arguably made matters worse for the Subcontractor, as the most rudimentary of details are taken to comply with the requirement to state 'the basis on which that sum is calculated'.

Once again it is irrelevant if the sum the payer thinks is due is zero; the notice must still be served. The parties can agree in the contract how long before the final date for payment a pay less notice must be given, but if they do not agree this, then the period in the Scheme for Construction Contracts applies (7 days before the final date for payment).

The right to suspend performance

If there is no valid notice to pay less given by the payer and the payment is not made fully by the final date for payment, then the payee is entitled to suspend performance of part or all of his obligations under the contract (see s112). Previously, the whole of the works had to be suspended, but now the payee can choose to suspend only part if necessary.

The payee has to give the payer 7 days' notice of his intention to suspend work stating the ground or grounds on which he is suspending. Once the payer makes the payment in full, the right to suspend ceases and the payee could be in breach of contract if he does not then resume work.

The payee is entitled to an extension to any contractual time limit whilst the payee's obligations are suspended. The LDEDC has also introduced the right for the payee to claim from the payer the reasonable costs and expenses incurred as a result of the suspension – this might include remobilisation costs, for example.

The Scheme for Construction Contracts

The Act provides that the relevant provisions of the Scheme will apply where a contract fails to comply with certain specified provisions of the Act.

Payment mechanism

The Scheme applies where the parties to a relevant construction contract fail to agree:

- The amount of any instalment or stage or periodic payment for any work under the contract, and/or

- The intervals at which, or circumstances in which, such payments become due under that contract.

If the parties do not provide an adequate mechanism to determine what payments are due and/or when they are due, the Scheme provides as follows:

Due date

A payment becomes due on the later of

- The expiry of 7 days following the 'Relevant Period' (if the contract does not specify a relevant period or one cannot be calculated by reference to the contract then this period is 28 days); or the making of a claim by the payee.

Final date for payment

If the parties have failed to provide for the final date for payment of sums due the Scheme states that the final date for the payer to make payment is 17 days after the payment due date.

Payment notices

Construction contracts must provide for notices of payment to be given. If they do not, then the Scheme states that the payer must give the payee notice no later than 5 days after the payment due date. The notice must state:

- The sum the payer considers to be due or to have been due at the payment due date; and
- The basis on which the sum is calculated.

It does not matter if the sum he considers due is zero, a notice must still be given. And although the Act itself allows parties to agree that the payee can give the notice, or that another third party specified in the contract can give the notice, the Scheme only refers to the payer giving this notice.

Paying less than the notified sum

If the payer intends to pay less than the sum notified in the payment notice, he must serve a pay less notice. If the parties have failed to agree the period in which that notice must be served the Scheme provides that the notice must be given no later than 7 days before the final date for payment.

Pay when paid

The Scheme will also automatically apply if the parties have stated that payments are conditional on the payer receiving payment from a third party (which is prohibited by the Act unless that party is insolvent) but have failed to provide any other terms for payment.

Your Subcontractors

It is vital to realise that the Act applies equally to your Subcontractors or Sub-Subcontractors, (even the labour only ones) and you must therefore take care to ensure your own contracts comply with the Act. Similarly, you must ensure that you comply with your obligations as regards payment and pay less notices.

Conclusion

Whilst the payment mechanisms introduced by the Act, and amended by the LDEDC might seem complicated, they have introduced vital provisions to protect you. Therefore, you should always be very wary when Contractors offer you their own terms and conditions, or seek to amend the Standard Form contracts such as JCT and NEC.

Familiarise yourself with the rules, and don't allow the Contractor to change them!

Adjudication under the Act

Where the Act applies (see the first chapter of this section), there is now a statutory right to refer any dispute or difference arising under the contract to adjudication. All 'construction contracts' as defined in the Act must contain an adjudication procedure. The basic requirements of the Act are as follows:

1. Either party can give notice of adjudication 'at any time'.

2. The contract must provide a timetable for appointment of an adjudicator and referral of the dispute within 7 days of the initial notice.

3. The adjudicator must reach a decision within 28 days of the referral (up to 42 days if the referring party agrees).

4. Period extended only if both parties agree, or with consent of referring party.

5. The adjudicator must act impartially.

6. The adjudicator may take the initiative in ascertaining the facts and the law.

7. The decision of the adjudicator must be stated to be 'binding until the dispute is finally determined by legal proceedings, by arbitration … or by agreement'.

8. The parties may agree to accept the adjudicator's decision as final.

9. The adjudicator can correct clerical or typographical errors in his/her decision.

10. The adjudicator is not liable for anything done or omitted unless the act or omission is in bad faith.

If the contract agreement fails to comply with any of the above basic requirements, then the entire provisions of the 'Scheme for Construction Contracts' shall apply. And you also need to bear in mind that the contract may contain express provisions which provide for adjudication, even if the Act does not apply.

What is Adjudication?

Adjudication is a statutory legal procedure by which any party, and particularly a Subcontractor, to a construction contract has the right to have a dispute decided by an adjudicator. It is intended to be a quick process and it can be cost-effective when handled properly.

It is normally used by Subcontractors to obtain payment, but most types of dispute can be adjudicated. However, some disputes are not suitable for adjudication, and some Contractors will not pay even when the adjudicator's decision goes against them. Therefore, the author strongly suggests that you take appropriate advice before setting off on the path to adjudication. It is very easy to get it wrong, and getting it wrong can have serious financial consequences!

What type of disputes can be adjudicated?

Please don't assume that you automatically have a dispute that can be adjudicated. For a dispute to be adjudicated, that dispute has to 'arise under the contract'.

First, this means that disputes relating to the formation of the contract are not ordinarily capable of resolution by adjudication.

Second, the 'dispute' may not yet have arisen, and adjudication should not be used to replace the contractual processes. In other words, if you haven't properly submitted your claim or followed the procedure set out in the subcontract, or have been asked for information which you haven't provided, you will not be able to adjudicate unless the other party agrees. Similarly, the other party may genuinely not understand your point of view. You need to take an objective view of that situation and if in any doubt take professional advice.

Third, a great many disputes can be resolved by dialogue and negotiation and without wanting to go soft on the unscrupulous Contractors, you should check that you have done all you can to resolve the subcontract dispute by discussion. This might require some tough talking, but one big advantage of negotiating a settlement is that you get certainty. In other words, if you agree a deal for £'x' then that is what you are going to get.

If you commence adjudication or any other form of formal dispute procedure set out in the subcontract, such as litigation or arbitration, you can never be certain of the outcome, no matter how strong a case you have. Look at it this way. When two professional and unbeaten boxers get into the ring, both think they are going to win, but as we know, that is impossible!

The kind of disputes that can be adjudicated include

- Non-payment of the Subcontractor
- Under-certification of the Subcontractor's account

- Failure to properly value subcontract variations
- Set-off from the Subcontractor's account
- Entitlement to extension of time
- Entitlement to loss and/or expense

Are there any special procedures?

The contract may contain provisions and rules for adjudication. It may incorporate a set of rules by reference, or it may be silent on the issue.

You need to be very careful that you understand the rules! Just like everything else in construction, some Contractors will bend the rules in their favour. The amount to which the rules might be drafted against you will vary from the inconvenient to the downright lethal. See the chapter on 'Tricks and Dodges'.

There are a number of organisations which have published their own Adjudication Rules and/or act as Adjudicator Nominating Bodies (ANBs). Here are a few of them:

> The Technology and Construction Solicitors' Association ('TeCSA')
>
> The Technology and Construction Bar Association (TECBAR)
>
> The Institution of Chemical Engineers (IChemE)
>
> The Chartered Institute of Arbitrators (CIArb)
>
> Chartered Institute of Arbitrators (Scotland) (CIArb-Scotland)
>
> Chartered Institute of Building
>
> Institution of Civil Engineers (ICE)
>
> RICS–Dispute Resolution Service (RICS–DRS)

Construction Act adjudication was originally intended to be a simple process by which disputes could be resolved. To a certain extent adjudication is a simple process, but it is nonetheless a legal process and not one to be undertaken lightly.

If you have no experience of legal process, then it is best to take professional advice. Bear in mind that the whole adjudication process takes place very quickly, so you need to act fast if you are on the receiving end of an Adjudication Notice!

Starting the process – the Adjudication Notice

Once you are satisfied that you have a dispute that can be adjudicated, the process is started by way of written notice (Adjudication Notice) to the other party.

The Adjudication Notice is a very important document in that it defines what matters the adjudicator has the jurisdiction to decide.

As a very minimum the notice must contain the following:

- Nature and brief description of the dispute
- When and where the dispute arose
- Nature of the redress being sought
- Names and addresses of the parties to the contract

The Adjudication Notice must be served on the other party before you approach an Adjudicator Nominating Body to appoint an adjudicator. The ANB will normally require a fee for them to appoint an adjudicator.

Once appointed, the adjudicator will write to both parties advising them of his/her appointment.

The Referral Notice

The Referral Notice must be issued to the adjudicator within 7 days of the issue of the Adjudication Notice. There is usually a fair amount of work in the preparation of the Referral Notice even in the simplest of disputes. It is for this reason that in the vast majority of cases the Referral Notice is prepared before the Adjudication Notice is served.

The 28-day period for the adjudicator to make his decision starts on the date the adjudicator receives the Referral Notice, provided of course that it has been served within 7 days of the Adjudication Notice.

Who pays for adjudication?

Although adjudication is generally inexpensive in comparison with arbitration or litigation, the process is not free and there are inevitably some costs that have to be paid. There are two elements to these costs: the fees of the adjudicator (together with those for any advice and assistance obtained by them) and the costs that you and the other party, as participants in the process, spend on your own legal, expert or commercial advice.

Who pays the adjudicator's costs?

Who pays the adjudicator's costs is one of those matters that depends upon the terms of the adjudication procedure. The Act requires that the adjudicator is entitled to decide who should pay the adjudicator's costs, as part of the decision, unless the parties have agreed otherwise after the notice of adjudication has been given.

Often, the adjudicator will decide that the party 'losing' overall must pay their costs. However, this is not always the case, and the adjudicator may take into account matters such as how each party has behaved, and whether each party has won on some issues. On the other hand, whatever the outcome of the decision, the adjudicator may simply apportion the fees equally between you and the other party.

This is not the end of the matter since both parties are jointly and severally responsible to the adjudicator for their fees. This means that if the other party does not pay, you will have to: if one of you defaults on payment or becomes insolvent, the adjudicator can legally demand those fees from the other, leaving that other party to recover from the defaulter. It is also worth remembering that the adjudicator is under a duty to avoid incurring unnecessary expense.

The adjudicator has obtained expert advice – do I have to pay for this too?

Provided the parties have been notified first, the adjudicator is entitled to appoint experts, assessors or legal advisers as required. Within the general requirement to avoid incurring unnecessary expense, the costs of any such external advice will form part of the adjudicator's costs. Similarly, the adjudicator may require tests or experiments to be carried out and the costs of these will also form part of the adjudicator's charges.

What will the adjudicator charge?

There is no set rate for an adjudicator; a range of hourly rates are charged. The total amount will depend upon the complexity of the issues and the length of time the adjudication takes. If the adjudicator was named in the contract or agreed by the parties, then the hourly rate should be agreed at that time. However, if the adjudicator is nominated by an Adjudicator Nominating Body, there is no opportunity for the parties to agree an hourly rate and the adjudicator must set a reasonable rate.

Can I ask the adjudicator to award me my costs?

Whether the adjudicator has the power to award the parties' costs (as opposed to the adjudicator's own costs) depends upon the terms of the adjudication procedure. Under the Scheme, the adjudicator does not have this power, and many, but not all, specially written procedures specifically provide that each party pays its own costs.

The parties can make a one-off agreement to allow the adjudicator to deal with the parties' costs; if, for instance, each party in its submission asks the adjudicator to allocate party costs, this will be construed as such an agreement, so be careful as to whether or not you are agreeing to the adjudicator having the power to deal with the parties' costs.

How does the process end?

It sometimes happens that the other party will seek to negotiate a settlement once the Notice of Adjudication has been issued. You should be alive to this possibility but should not expect it to happen. Nor should you become distracted from the adjudication process itself. It is often the case that the Notice of Adjudication brings about a hardening of attitude from the other party who will then fight tooth and nail to defend their position.

If the process runs all the way through, it will result in the adjudicator's decision. This will be published to the parties and it will set out the decision in respect of the matters referred to him/her. It will also set out the date by which any monies to be paid by one party to the other must be paid, and it will set out his/her fees and the proportion to be paid by each party.

If you are the Referring Party and you obtain a monetary decision in your favour, then you should contact the Responding Party to establish how they intend to make payment, e.g. BACS transfer, collect a cheque, etc.

Is the adjudicator's decision legally enforceable?

If you win and the losing party refuses to pay, then it is a matter for you to enforce the decision in the courts, and this is a job for a solicitor with specialist construction expertise, as it should be a relatively quick process.

The adjudicator's decision is binding on the parties until the dispute is finally decided by arbitration, litigation or agreement. Because it is binding, the courts will enforce the decision in summary proceedings. In order to do this, it is necessary to issue a claim form and an application for Summary Judgment in the Technology & Construction Court.

Courts aim to provide a hearing date within 28 days of the issue of the claim form, and this is rightly a tight timescale for the Defendant to serve its evidence in response, setting out his reasons for non-payment and for the Claimant to respond prior to the hearing date. Ordinarily, provided the Judge is satisfied that the adjudicator's decision is valid, and that the Claimant is entitled to enforce that decision, then Judgment will be given in the Claimant's favour, together with an award of costs.

Challenging an adjudicator's decision

Adjudication is sometimes referred to as a form of 'rough and ready' justice. Not surprisingly therefore, there are occasions when the decision reached, or the way it has been reached, give rise to the responding party wanting to challenge the decision.

Challenging an adjudication decision can only be successful for certain specific technical reasons. Simply having received what appears to be the 'wrong decision' is not sufficient grounds for a challenge. As a form of 'rough and ready justice', adjudication has worked well.

Adjudicators have quite sweeping powers of investigation, but they must act within certain defined limits, and if they exceed those limits then their decision could be unenforceable.

Adjudicators have to act within the jurisdiction which has been given to them, and in accordance with the rules of natural justice. If they do not, then the courts may refuse to enforce their decision.

What is the Scheme for Construction Contracts?

The Act provides, at s108 that a 'construction contract' must embody eight principles concerning the right to adjudication. If the contract does not comply with Section 108, then any adjudication provisions in the contract (except for the naming of an adjudicator or adjudicator nominating body) are ignored and the Scheme applies.

The Scheme for Construction Contracts is a set of rules which includes all eight principles and will automatically apply. The Standard forms of contract generally include all of the principles, but be extremely careful that the contract does not try to amend the Scheme.

Is adjudication the only way to resolve the dispute?

Most certainly not! In fact, it may be that adjudication is definitely not the best way to resolve the dispute. There are potentially various ways to resolve the dispute ranging from informal negotiation to litigation.

If your own efforts to settle the matter have failed, the sooner you take professional advice the better.

Conclusion

Adjudication is designed to be a straightforward process to enable disputes to be resolved quickly and inexpensively. In some cases, it may be unnecessary for you to incur the cost of obtaining professional assistance from lawyers, consultants or other specialists. However, adjudication is a serious process and mistakes can be very costly.

The proper preparation and presentation of your written case with supporting evidence to the adjudicator will invariably be a major factor in determining the success or failure of your arguments. Don't forget that the adjudicator only has a short time in which to consider the arguments put forward by both parties before reaching a decision.

If you are considering adjudication, you should always take professional advice before proceeding.

Handled well, adjudication is an effective means of resolving disputes.

Tricks and dodges of the Construction Act

Unfortunately, despite the 2011 amendments to the Act, and the expressions of 'fair treatment' and 'strong long-term relationships with our suppliers', that you find on all the big Contractor's web sites, the fact of the matter is that the 'loopholes' in the Act have been and continue to be exploited by many Contractors. These same Contractors evade and abuse the spirit, and often the letter, of the Act. Some examples are listed below. Look out for them at every stage of the enquiry to contract process, and don't accept them. We vet lots of Contractors terms and conditions, and you will be pleased to know that they will agree to changes.

Payment

Extended payment periods

Now that 'pay when paid' is outlawed, many Contractors are imposing greatly lengthened payment periods, by either stretching the number of days until the 'Due Date' and/or between the 'Due Date' and the 'Final Date' for payment. You would do well to consider that where the Act says, 'the parties are free to agree', it actually means 'free to get screwed!'

Payless Notice

The Act does not prescribe a minimum period of notice. The majority of Contractors state that a Payless Notice can be issued as late as only 1 day prior to the Final Date for payment. Clearly, this is totally inadequate and could have appalling effects on your cash flow. It means that if you accept say 60 days between the Due Date and the Final Date, the Contractor can tell you in his Payment Notice that he is going to pay you £100,000, and then on day 59, he can issue a Payless Notice stating that he is only going to pay you £20,000!

Cross set-off

Many Contractors are including 'cross set-off' from other subcontracts. Some even extend this to other companies in a group.

Future set-off

Some Contractors are imposing the power to set off 'future costs' which are 'likely to be suffered or incurred', based merely on their own estimates. This can easily be abused, and trumped up costs made to look genuine.

Notice of suspension

The Act says 'not less than 7 days' notice to be given by the Subcontractor in the event of non-payment of sums which have been notified by the Contractor as due. Some unscrupulous Contractors are imposing greatly lengthened periods or are introducing other onerous requirements that have to be complied with before your suspension notice can be 'valid'.

Pay when certified

With 'pay when paid' outlawed, except where the Client is insolvent, some Contractors were (and some still are) fixing the due date as the date of certification by the architect under the Main Contract (i.e. 'pay when certified'). You would have no idea when this will happen, or if certification is withheld due to some default of the Contractor. Fortunately, this provision is now caught by the 2011 amendments to the Act. It is no longer enforceable and the Scheme will apply.

Client or Employer's insolvency

Most subcontracts now have a clause, which takes advantage of the 'pay when paid' loophole in the event that the party paying the Contractor becomes insolvent. It is vital that you either get the clause struck out or thoroughly check that party's credit rating!

Valid applications

Beware subcontract conditions that stipulate rigorous requirements to be met by the Subcontractor when submitting your interim application (e.g. full details, proof of ownership of all materials and equipment and fully priced dayworks and variations).

If these requirements are not met, then the application does not qualify as a 'valid application' and therefore no payment becomes due. You must ensure that you comply or risk the consequences.

Adjudication

'Dissatisfaction procedure'

A very common ploy is to introduce a 'dissatisfaction procedure' before you can give notice of adjudication. Some even fix a prescribed period (e.g. 2 months). However, you will be pleased to know that the Act provides that a party to a construction contract **has the right at any time** to refer to adjudication a dispute arising under the contract. If this right is not there, then the Scheme applies so there is no bar to getting on with it.

Joining-in clause

This is a clause that obliges you to join in with the Contractor against the Employer, if the adjudication matters are related. This is another device intended to thwart your right to prompt adjudication, and/or get you to put their case together for them.

'Final and binding'

Some Contractors are imposing a 'final and binding' clause in respect of their own decisions, to prevent the adjudicator from reopening those decisions (e.g. extension of time and set-off). In practice, it means that a Contractor could refuse an extension of time and/or set off vast sums and you being unable to refer this dispute to adjudication.

Enforcement of adjudicator's decision

Some Contractors will not pay even if the adjudicator decides matters in your favour, and you will have no alternative but to go to Court to enforce the adjudicator's decision. The best course of action is always to take appropriate professional advice before commencing adjudication.

'Stakeholder' clause

Some Contractors have introduced a 'stakeholder' clause by which the sums arising from an adjudicator's decision are placed with a stakeholder, rather than paid to you.

These types of provisions are usually accompanied by clauses dressed up to make them look like they are 'reasonable', but rest assured they are anything but. And nor are the Contractors that include such clauses in their contracts!

Costs of adjudication – adjudicator's fees

Who pays the adjudicator's costs depends upon the terms of the adjudication procedure. The Act requires that the adjudicator is entitled to decide who should pay his/her fees, unless the parties have agreed otherwise after the notice of adjudication has been given. Often, the adjudicator will decide that the party 'losing' overall must pay their costs. However, this is not always the case and the adjudicator may take into account matters such as how each party has behaved, and whether each party has won on some issues.

Costs of adjudication – the parties' costs

Some Contractors are still stipulating that any party who gives a notice of adjudication will automatically pick up the full costs of the adjudication, including the adjudicator's fees and the Contractor's costs. This flagrant abuse of the Act effectively made adjudication a 'non-starter' in most cases.

Whilst this was grossly unfair, it was deemed to be permitted under the old Act. However, case law and the new Act have attempted to make such provisions void; it is debatable whether this has been 100% successful, and it may still be possible to draft Act compliant terms which makes one party liable for the other party's costs.

If both parties agree, the adjudicator is able to decide that one party will bear the other's costs in relation to success, so be careful what you sign up to, and take professional advice!

Costs of adjudication – payment on account of adjudicator's fees

Some onerous contracts prescribe a sum of money, which you must pay on account of the adjudicator's fees (a sum of £10,000 is commonly inserted). This 'down payment' must be made before you can give notice of adjudication, or is required at some stage during the process. Clearly, this is a deterrent to many smaller Subcontractors and a gross abuse which could be held to be contrary to the Act and your **right at any time** to refer to adjudication a dispute arising under the contract. Again, be careful what you sign up to, and take professional advice!

Conclusion

Adjudication is intended to be a quick and economical way to resolve disputes. So, in order to avoid problems with the process, my advice to all Subcontractors can be summed up as follows:

- **You must vet the contract terms very carefully**
- **If you don't understand the terms, get professional advice**
- **Take professional advice BEFORE commencing adjudication**

One last thing. In the Third edition, Jack asked that *'When you come across these dirty tricks and dodges, don't keep it a secret. Make sure word gets around'.* And he also asked you to remember that – 'For evil to triumph, it is only necessary for good men to do nothing'.

Were he alive today Jack would be dismayed at just how little things have changed, and that such tricks and dodges are still rife!

But one thing that has changed for the better is the ability to spread the word about unfair practices via social media and email and our #buildgate campaign. So please, whenever you see onerous provisions, get in touch and we will spread the word for the good of all Subcontractors, and we can of course do that anonymously.

> **Watch out for those dirty tricks, and take action to avoid them!**

The Construction Act – some questions and answers

Experience indicates that the same questions and problems crop up time and time again. Here are a few of the more common queries and the suggested answers. These are merely examples, and as each situation turns on its own special circumstances and wording, you are strongly advised to seek professional advice in all such matters.

Payment

What if the Client's or Contractor's subcontract terms take no account of the minimum requirements of the Act (e.g. payment by instalments, adequate mechanism for valuing and prior notices of sums due and/or withholding)?

Despite the Act having been in force since 1 May 1998, we constantly see contracts that pay scant regard to its provisions. Indeed, the fact that the Act applies to them still comes as a complete surprise to many parties! And they are equally surprised to find that if any one or more of the minimum requirements are not met, the relevant part or parts of the Scheme come into operation as a 'default' mechanism.

Very often engaging a professional to write to the other party pointing out the error of their ways is sufficient to get matters resolved.

What if the Contract states 'pay when paid'.

'Pay when paid' is made 'ineffective' by the Act, unless there is a clause relating to insolvency of a third party upon whom payment depends. Do not accept excuses such as 'I am still waiting for the Client to pay me for the last application!'

The Contractor just sends me a cheque with no details of how the figure is built up.

The Act requires that you be given prior written notice of sums due and the basis of their calculation. You are entitled to insist that the Contractor complies, but if the contract provides for interim applications, then it is entirely likely that your application has become the 'default' payment notice. If there is a significant difference between your own application and the Contractor's cheque, it is even more essential that he provides you with his own valuation and that you take action to resolve the matter.

The Contractor sends me a notice showing me his assessment of my monthly valuation, but when I receive the cheque, he has deducted a lump sum for 'contra charges', with no prior notice or explanation.

The Contractor is ignoring the Act. Once he has notified you of the sum due, he must make payment of that sum, unless he then gives you a Pay Less Notice setting out the basis of his calculation. Engaging a professional to point out to the Contractor that an adjudicator would instruct him to make payment in full may be sufficient to resolve the matter.

In this instance you are entitled to issue a notice of suspension.

The Act gives me the right of suspension by not less than 7 days' notice in the event of late payment. If the Contractor's notified valuation is less than my application, can I suspend my works?

No. The right of suspension is related to late payment of the sum which has been notified as due. There is no right of suspension merely because of a difference of opinion regarding your valuation. What you have is a dispute about 'valuation' not 'payment'. They are separate issues, but the valuation can be referred to adjudication if appropriate. Again you should take professional advice before even threatening to suspend work.

My subcontract is for electrical works to a new factory. My Client has given me a purchase order more suitable for manufacturing and supplies. The payment terms are worded accordingly. Am I entitled to ask for payment terms which comply with the Act?

If this is a 'construction contract' under the Act, then the contract must include payment terms which comply with the Act. If it does not, then the terms of the Scheme will be implied into the contract in respect of each requirement that has been ignored in the order.

Adjudication

The Contractor's terms make no mention whatever regarding 'adjudication'. Where do I stand?

The Act states that if the contract conditions fail to comply with any one of the basic requirements, then the entire provisions of the Scheme will kick in by default.

I have a dispute with a Contractor regarding the value of my variations. Yesterday's meeting was the last straw, and when I got back to my office,

I issued a notice of adjudication. Now I wonder if I might have been hasty. What do you think?

If you 'flew off the handle' without first preparing a separate referral notice, with all the details of your dispute, etc., then you could be in trouble. The adjudicator will be on board in a matter of days and will be asking to see your referral notice. If you cannot comply, then your adjudication will collapse at the first hurdle! Always remember, before you even mention the word 'adjudication' to the other party, get professional advice to ensure that you are properly prepared.

I have a very complicated dispute with the Contractor regarding my account and a loss and expense claim. The job overran by 10 months, and there are numerous files packed with the correspondence which we have exchanged over a long period. Is this a suitable case to refer to adjudication?

Perhaps not. Adjudication was originally intended to be a means of obtaining 'justice' within a matter of a few weeks. It is very difficult for an adjudicator to reach a sound decision on large, complex disputes in the limited time available.

Potentially, you could identify certain fundamental issues and referring those issues to the adjudicator. This could be helpful to the parties and gives the adjudicator adequate time to reach a carefully considered judgement, which may well aid the parties in reaching a settlement.

Whatever else you do, it sounds like you should definitely take professional advice!

What kind of disputes are most suited to adjudication?

Adjudication is ideal for settling cases of blatant disregard of the Act by the paying party. In general terms, the simpler the issues, the more suitable the case is for adjudication. Adjudicators commonly charge a minimum of £250 per hour (at 2017 prices) plus expenses. I was appointed by a Client Subcontractor after he had referred a matter of non-payment and a complex mishmash of variations to an adjudicator, with a dispute in the order of £144,000.

He lost the adjudication and had to pay £10,000 fees to the adjudicator, plus his own costs. It cost him a small fortune to get absolutely nothing out of a £144,000 dispute.

Fortunately, by setting about the matter in the right way, we were able to get the matter resolved, and a further and substantial payment was ultimately received. So, it is essential that you take professional advice and weigh up all the pros and cons before rushing into something you may live to regret.

Know your rights – understand the Construction Act!

Section 8

The JCT and NEC subcontracts

JCT 2016 – what's changed?

In 2016, the Joint Contracts Tribunal (JCT) rolled out the latest major update to its family of standard form construction contracts. From the Subcontractors point of view, it is essential to get to grips with these changes, and particularly how Contractors will seek to amend them.

You should not be surprised to know that the contracts we have vetted thus far show that most Contractors will be keen to avoid passing on any benefits that the new provisions were intended to introduce!

Overview

A number of new features are common to all the 2016 family of contracts. These include

- Incorporation of the provisions of the JCT Public Sector Supplement 2011 that relate to fair payment, transparency and business information modelling.

- Adjustments to reflect the Construction (Design and Management) Regulations 2015 and the Public Contracts Regulations 2015.

- Amendments to the works to existing structures insurance provisions (Option C) to make them more flexible.

- Revision and 'simplification' of the Section 4 payment provisions, including

  - Establishing 'interim valuation dates' (IVDs) that apply throughout the supply chain.

  - Increasing flexibility in the fluctuations provisions.

  - Consolidating the notice provisions required by statute.

- Introduction of a procedure for prompt assessment of loss and expense claims.

- Inclusion of provisions for the grant of performance bonds and parent company guarantees.
- Extending the option to Subcontractors of using third-party rights (TPR) instead of collateral warranties.
- Incorporation of the JCT 2012 Named Specialist Update.

On the face of it, the general position adopted by the JCT in relation to risk allocation has not changed. Because of that, the likelihood is that developers and Contractors will continue to negotiate amendments to the standard forms to reflect their own approach to risk. And as Contractors like to pass on as much risk as possible to their Subcontractors, it is essential that you familiarise yourself with the new procedural requirements of the contracts, in particular those that deal with payment and loss and expense claims.

Payment terms

The most significant change to all the 2016 Contracts relates to the restructuring and simplification of the payment provisions. The main points to note are as follows:

- IVDs have been introduced, establishing a common valuation date. This is designed to speed up payments throughout the supply chain, in line with the government's Construction Supply Chain Payment Charter.
- The contracts no longer distinguish between interim payments due before practical completion (PC) and after PC. Accordingly, the period between interim certificates issued post-PC has reduced from 2 months to 1 month.
- Notified sums (plus interest) are now automatically recoverable as debts.
- Common valuation dates.
- The unamended 2016 Contracts establish the following multi-tiered payment regime:
  - The Main Contract states an IVD, which applies on the same day each month (or nearest business day). The same date is written into all the subcontracts and sub-subcontracts, which themselves need to use the appropriate JCT 2016 form.
  - The first IVD must now be stipulated in the contract particulars (instead of the due date for payment). If the contract particulars are not completed, then the first IVD is 1 month after the works commencement date and then at monthly intervals.

The aim of the amendments throughout the contractual chain is to allow payment to flow down to all tiers within 31 days of the IVD. Accordingly, the JCT subcontracts set the due date 12 days after the IVD, so that the main contractor is put in funds to pay the Subcontractor with a 5-day margin. The sub-subcontract sets the due date 17 days after the IVD, putting the Subcontractor in funds to pay its Sub-Subcontractor with a 5-day margin.

Call me cynical but I expect most Contractors to amend these standard forms, so that they don't have to comply with the multi-tiered structure of the new payment regime, and you should make it clear (especially on publicly funded projects) that you know what should be happening and ask the Contractor to comply!

Collateral warranties/TPR

The streetwise subbie needs to check out item 6 of the Sub-Contract Particulars and Clause 2.26 which deal with Purchaser and Tenants Rights.

The amendments now extend the optional provisions for collateral warranties from Subcontractors to include TPR, or by way of collateral warranties.

Whatever else you do, you must ensure that you understand the obligations that you are entering into. They could well be onerous. Take a look at Section 1 'Collateral warranties'.

Loss and expense

The changes to the loss and expense mechanism in Section 4 of the DBC and SBC need to be considered carefully. The JCT guidance confirms that the intention is to improve the timing and certainty of loss and expense claims.

In the 2011 editions, the Subcontractor must notify:

> …as soon as it has become, or should reasonably have become, apparent to him that the regular progress has been or is likely to be affected.

Under the new provisions, the duty kicks in a little earlier. Now the Subcontractor must notify:

> …as soon as the likely effect of a Relevant Sub-Contract Matter on regular progress or the likely nature and extent of any loss and/or expense arising from a deferment of possession becomes (or should have become) reasonably apparent to him.

Previously, the Subcontractor had to supply its supporting information 'upon request'. Now it should be submitted with the notice or 'as soon as reasonably practicable thereafter'.

In addition, the Subcontractor is required to provide monthly updates 'until all information reasonably necessary to allow ascertainment of the total amount of such loss and expense has been supplied'.

Time constraints are also placed on the Contractor who must now assess the initial claim within 42 days of receipt and within 28 days of each subsequent update. Previously, there was no time limit.

As before the cynic in me thinks that these improvements will definitely get the chop by most Contractors. Stand up for your rights and don't accept onerous amendments!

Will Contractors remove the benefits of the 2016 version of the JCT contracts?

The standard JCT building subcontracts SBCSub/C, SBCSub/D/C and DBSub/C

In late 2005, the JCT published its new replacements for our old friends, DOM/1 and DOM/2. The new forms follow the principle of having separate Articles of Agreement and Terms and Conditions documents. The forms were revised in 2011 to take account of the amendments to the Construction Act and as set out in the previous chapter, they were revised again in 2016 to take account of various issues (see JCT 2016 what's changed?).

As the JCT Sub-Contracts (and particularly the 2011 version) have been in widespread use for so long, it may take some time for the 2016 version to be widely adopted. But, whichever version you are offered, it is worth considering some of the key provisions, so that you can be alive to any onerous amendments.

Comprise two documents

Articles of agreement

Sub-Contract conditions

Articles of agreement

This document is where the job specific details are set out. The Contractor may incorporate these into a document containing his amendments to the Standard Form contract. They deal with

- Date contract formed
- Details of the parties to the contract

- Recitals
- Articles 1–6
  - 1 The documents forming the Sub-Contract
  - 2 The Sub-Contractors Obligations
  - 3A Sub-Contract Sum – Lump sum
  - 3B Sub-Contract Tender Sum – Re-measurable
  - 4 Adjudication
  - 5 and 6 Arbitration or Litigation – An option see the relevant footnote
- Sub-Contract Particulars
  - Conditions – the relevant Sub-Contract conditions
  - Arbitration – whether it applies
  - Base Date – relevant to fluctuating price contracts
  - Address for Notices
  - Programme
  - Attendance
  - Interim Payments – Clause 4.9.1 or 4.9.2 for 2011, or IVD for 2016 version
  - Listed items – materials and goods paid for off-site
  - Retention
  - Retention Bond
  - Fluctuations – now eliminated from 2016 version
  - Dayworks
  - Insurance
  - Incorporation of the Sub-Contract Works
  - Settlement of Disputes
  - Numbered documents
- Attestation – where the document is executed
- Schedule of Main Contract Information
  - Main Contract
  - Construction Phase Plan
  - Programme Information

Sub-Contract conditions

The Sub-Contract conditions set out the rules of the game:

1. Definitions and interpretation
2. Carrying out the Sub-Contract works
3. Control of the Sub-Contract Works
4. Payment
5. Valuation of Work and Variations
6. Injury Damage and Insurance
7. Termination
8. Settlement of Disputes

Schedules 1–7 in the 2016 version replace Supplemental Provisions, Variation Quotation, Bonds, Fluctuations, Code of Practice, etc. that were in the 2011 version.

Onerous amendments

Unfortunately, many Contractors are likely to amend the standard JCT forms, just as they did the old one. I could probably fill a whole book with the onerous terms I have come across in practice. Onerous terms usually occur in the Client's or Contractor's own 'non-standard' documents but can also arise as amendments to Standard Form contracts. These terms are deliberately designed to reduce your entitlements and increase your obligations. They will certainly make your life more difficult; they could end up costing you a substantial amount of money, and in the worst-case scenario, they may even cost you your business!

The simplest way to put onerous terms into context is to ask yourself this question; 'Why, when there are all manner of Standard Form contracts around such as JCT, NEC and MF/1, does anyone need to produce their own terms?' It isn't for your benefit!

The streetwise subbie is advised to identify these changes at enquiry stage, or during the process of getting in to contract, and seek to restore the standard terms.

Obligations regarding time

A phrase frequently found in non-standard contracts is 'time shall be of the essence'. That means that obligations regarding time are fundamental terms of the contract, and if they are breached by the Subcontractor, the Contractor has the right not only to claim damages but also to treat the subcontract as being at an end.

You should also beware of general obligations requiring you to comply with the Contractor's programme and all of his directions regarding the order of the work, etc. Such clauses mean that you will be at the beck and call of the Contractor, and you will find it almost impossible to claim for delay and disruption.

Design

Ensure you know exactly what your design obligations are. And exactly what design documents you are working to. A one-line amendment, giving precedence to the 'Room Data Sheets', cost one of my Clients £30,000!

Do not accept 'fitness for purpose' obligations or allow these to arise by default, e.g. where the contract does not expressly limit your obligations to 'reasonable skill and care'. Fitness for purpose is onerous and will void your PI cover!

Compliance with the Main Contract

If you don't know what it says, how on earth can you price the risk of complying with it? Don't accept provisions that make you responsible for everything that's in the Main Contract. The subcontract needs to accurately reflect what you have priced to do.

Documents and errors and discrepancies

A common amendment is to make you responsible for all errors and discrepancies in the documents, even those provided by the Contractor! They do this by requiring you to point out any errors or discrepancies and then instructing you which requirement applies, but such instructions are to be carried out without any adjustment to the sub-contract price or the completion date.

Payment

The first common misconception relates to cash discount. It used to be the case that 2.5% cash discount was deductible only if the Main Contractor payed within the period required by the contract. It is nothing more and nothing less than an incentive to prompt payment. Unfortunately, the Contractor now wants to take discount irrespective of when he pays. If the Contractor wants a discount, make sure it is conditional on prompt payment.

The second and simplest amendment is simply to lengthen the payment periods. They will do this by increasing the period up to the Due Date and/or the period from the Due Date to the Final Date for payment, or both!

This is combined with a very short period (usually 1 day) prior to the Final Date for payment to issue the Payless Notice. A very deadly combination.

Retention

Check the retention periods carefully or it might be years before you get your retention money back!

Set-off

The Construction Act does not prohibit set-off against payments, but a Pay Less Notice must be given detailing the amount to be paid and how it is calculated. This is explained more fully in Section 7, but watch out for linking set-off to other contracts and 'anticipated' or future costs.

Insurance, protection and damage

Contractors may attempt to reduce their insurance premiums by making the Subcontractor responsible for damage to the subcontract works. Even that caused by his own negligence. Under a standard form, the Subcontractor is protected by the Contractor's policy if the damage is caused by one of the 'Specified Perils' which include fire, explosion and water damage, even if the Specified Peril is caused by the Subcontractor's negligence.

As your insurance policies will normally be worded to cater for the position under the standard forms, it is vital that such matters are picked up at the tender stage, or you could discover you are responsible but are uninsured.

Attendance

The provision of appropriate attendances by the Contractor is something that must never be assumed. Even under the standard forms of subcontract, it is vital that you have identified at tender stage items of special attendance you will need, who will be responsible for providing them, and that this is properly incorporated in the subcontract.

Under non-standard forms, items that are normally considered to be general attendances may not be provided, such as hoisting, storage space, electricity, water, messrooms, sanitary accommodation and welfare facilities.

Dispute resolution

Provisions may be inserted to attempt to restrict your entitlement to seek a legal remedy to any dispute which cannot be resolved by negotiation.

Conclusion

Watch out for onerous amendments to the standard form. Don't just accept what's put in front of you. Negotiate! And if you're not sure what that wording means, take professional advice.

Watch out for onerous amendments to the standard subcontract.

Have you met the NEC subcontract?

Have you met the New Engineering Contract (NEC), or to give it its correct title the Engineering and Construction Subcontract? The NEC (as it usually known) has now been in use for over 20 years and is ever more widely used. Although NEC contracts are published by the ICE, they are not in any way restricted to civil engineering. Instead they can be used for defining and then managing any project or service in the construction programmes of the hospital and education sectors, the water treatment industry, road construction and numerous other areas. Sooner or later you will come face to face. And when you do, you will need to be ready. Ready, because it is very different to the traditional approach under the JCT-type contracts. And ready to comply with its stringent demands.

If you do not want to get ready for the contract, then get ready for financial disaster!

The NEC contract was devised by the Institution of Civil Engineers, and it was first published in 1993. The second edition was published in 1995 (retitled 'The Engineering and Construction Contract'), and July 2005 saw the publication of an amended and expanded suite of documents, entitled NEC3. The most recent edition in being the NEC/4 which was published in 2017.

The governing philosophy of the NEC is stated to be '**mutual trust and co-operation**'. Indeed, the very first Clause states

> 10.1 The Contractor and the Subcontractor shall act as stated in this subcontract and in a spirit of mutual trust and co-operation.

However, the Subcontractor would be well advised not to overdo his reliance on 'trust' or 'co-operation'. The Contractors in almost every NEC-related dispute I have been involved with have shown a remarkable ignorance of this requirement!

The authors claim that the documents are written in 'ordinary language', with no legal jargon. Interestingly, many people, including lawyers and judges, actually find it harder to understand than the more traditional contracts.

The NEC is a manual of management procedures, not just a contractual document. It is definitely not something you lock in the safe and look at only if you have a problem.

The stringent rules must be followed at all times or risk dire consequences. Indeed, it is a working document full of procedures that should be used on a daily basis.

There are six main options A–F, plus a range of secondary options, any combination of which can be selected by the Client or Contractor as his chosen procurement route. The most common formats are Option A ('Priced contract with activity schedule'), Option B ('Priced contract with bill of quantities') and Option C ('Target contract with activity schedule'). The secondary options include such elements as advance payments, retentions, sectional completion, limitation of design responsibility, fluctuations, bonus for early completion, delay damages, etc.

There is '**contract data**' that acts as a highly detailed 'appendix' for the parties to enter specific information in the tender document (e.g. price, dates, periods, parties and functions).

Three key aspects of the NEC

Subcontractor's programme

- To be provided and updated at stated intervals as contract data or when instructed.
- Twenty-five per cent deduction from payment as financial penalty for failure to provide.
- Highly detailed. Shows key dates, order, timing, method statement for each operation and level of resources.
- Programme is used to assess Compensation Events.
- Is intended to be the 'driving engine' of the contract.

Compensation Events

- This is the only mechanism for additional payments, 'claims', variations, changes to completion date, etc.
- The qualifying events are listed in the contract, and it is a generously long list.
- The basis is actual cost and/or forecast of actual cost, plus a fee percentage (as contract data).
- No provisions whatever for 'top-ups' or residual 'end of job claims'.
- Written notice is required within tight timescale or entitlements may be lost.
- If the Contractor neglects to reply, then the Subcontractor may notify him accordingly and if the Contractor fails to reply again, the Subcontractor's notice is deemed as accepted.

- Quotation and forecast of programme effect is required for all Compensation Events.
- The idea is to 'value changes, delays and disruptions as the job proceeds'.

Early warning procedure

- The Contractor and Subcontractor each have a duty to give a written early warning as soon as they become aware of any matter which could increase prices, delay completion or impair performance of the works.
- Each party may instruct the other to attend a risk reduction meeting, along with relevant other parties.
- The parties must co-operate in making and considering proposals to overcome or minimise the problem.
- The Contractor then records in the risk register the proposals and the decisions taken and issues the same.
- If this procedure results in a change to the Works Information, then the Contractor must issue a formal instruction.
- If the Subcontractor fails to give an early warning notice when he should have done, then any Compensation Event is priced as though he had done so (in which case the Contractor would have been able to take action to avoid or reduce the problem).

Some other key features

The authors have chosen to implement the admirable objectives by means of very strict rules, procedures and time limits.

'Project Manager' (PM) to manage the job on behalf of the Client, so that his decisions 'reflect the Client's business objectives'. This is a dramatic change in traditional arrangements. Clearly, the PM is not required to act in any quasi-arbitrator role. If the Contractor disagrees with the PM's decision, then his redress is by reference to the Adjudicator.

Disputes procedure via Option W1 (in cases where the Construction Act does not apply) and Option W2 (where the Act does apply). This reflects the position in the process industries, much of which is excluded from the Act.

Subcontracting under the NEC

The use of the standard NEC subcontract is not mandatory. However, if the Contractor does not intend to use the standard form, he is obliged to submit the proposed conditions to the PM for approval. Presumably, the intention is to dissuade the Contractor from imposing his own onerous, in-house terms, although I have not seen much evidence of that!

'*Option*' for the subcontract does not have to be the same as that of the Main Contract. This allows flexibility, so that a Contractor under Option A (Priced contract with activity schedule) could, for example, engage the Subcontractor under Option B (Priced subcontract with bill of quantities).

'*Accepted programme*' is to be identified in the subcontract data. If not, it is to be submitted within a stated period, and it must be revised at pre-stated intervals and when instructed by the Contractor. The revised programmes must show full details of progress and the delay effects of Compensation Events and Early Warning matters. The vital importance of the programme is shown by a punitive financial provision for one quarter of the value of work done to be withheld until a first programme has been submitted.

'*Sub-subcontracting*' terms and conditions, if not in the standard form, must be submitted to the Contractor for approval. Again, a statement as to 'mutual trust and co-operation' is compulsory.

'*Delay damages*' can be inserted in the subcontract data at a daily rate. If so, they must represent a genuine pre-estimate of the Contractor's likely losses in the event of delayed completion by the Subcontractor. They effectively become 'liquidated' damages.

You will by now have realised that the NEC contract is quite different from the JCT forms to which we are accustomed. In the next chapter, we shall highlight some of the problems and pitfalls of which to be aware.

The NEC is very different from other standard forms!

NEC problems and pitfalls

The New Engineering Contract (NEC) contract brings many benefits to all involved parties, providing the rules are closely followed by all concerned. However, there are some significant problems awaiting the unwary.

Whenever I am asked about the NEC contract, I advise that there is nothing wrong with the NEC contract itself, but, in practical terms, there are three significant problems for Subcontractors:

1. It is rarely set up properly.
2. It is not understood by either party.
3. It is not administered properly.

These three issues inevitably come back to haunt Subcontractors in one way or another! So, on that cheery note, I will touch on a few of the other problems as follows:

Starting out on the right foot

One of the most common causes of NEC problems arises when the Subcontractor commences his works in the absence of properly agreed and detailed subcontract data. The NEC depends on all the various elements of the subcontract data being completed and agreed. If the parties commence an NEC relationship without having even agreed the most basic details (and that is often the case!), it is no wonder that the result is chaos and financial agony for the Subcontractor. So, it is vital that every aspect of the subcontract is agreed and put in writing before work commences. The NEC is no place for 'letters of intent', vague or otherwise.

The dreaded 'Z' clauses

Why are Z clauses problematic? The biggest problem with Z clauses, and in fact any extra clause inserted into any contract, is that they can change the nature of the contract. Z clauses are often used by unscrupulous Contractors to make life difficult for their Subcontractors.

Indeed, in its worst case I have seen over 100 pages of Z clause which made a complete mockery of the aims of the NEC. If you are struggling to understand the methodology of an unamended NEC contract, just think how much more difficult it is going to be to understand one that has onerous amendments!

The importance of the programme

Practitioners will argue that the programme is the most important document in an NEC contract. After all the deduction of 25% of all payments until such times as an NEC compliant programme is submitted should be telling you something!

To be compliant the programme must meet the requirements of clause 31.2. If you compare these with those of a JCT contract you might consider them to be onerous, or at least very demanding. On the positive side, a properly detailed NEC programme will be invaluable in ensuring that you obtain your entitlement under the contract and can also be used for managing the works on site.

However, the majority of Contractors and Subcontractors do not comply with the requirements and this can soon result in a loss of entitlement. If you do not have the resources to carry out programming as required by NEC contracts, it is important that you either retrain your staff, or employ the services of someone who fully understands the NEC requirements and processes.

Proliferation of notices, etc.

How will the Client or Contractor react? The contract positively insists that the Subcontractor shall provide timely notices of all problems, as a matter of management procedure. Having chosen the NEC contract, are the Client/PM and Contractor prepared for the very substantial increase in the number of 'notices' from the Subcontractor? Or will they regard these notices as 'confrontation'? From personal experience, I can confirm that this is often the case. So, it is a good idea for the Subcontractor to seek a pre-contract discussion at senior level, in order to establish the lines of communication, and most importantly, to let the Contractor know you WILL be complying with your obligations! This simple measure could save a lot of misunderstanding and bad feeling later.

No 'rolled-up' or 'end of job' claims for delay and disruption

The NEC is not like so many other contracts, where the norm is for final accounts to drag on for years and claims for loss and expense may be submitted months after the end of the job. On the contrary, the NEC is based strictly on a regime whereby all matters of time and money **must** be notified and resolved as the works proceed. There is no provision whatever for 'end of job' claims and the like!

All too many Subcontractors allow dissatisfactions to fester and to go unreported (e.g. response, or lack of any response, to their early warning notices and compensation events). Indeed, I have come across Subcontractors who told me that they submitted dozens of early warning notices without response, and similarly with regard to compensation events!

The NEC is so written that in the event of the rules being ignored, then it will be the Subcontractor who pays the penalty in terms of loss of his entitlements. The wise Subcontractor will therefore maintain a systematic 'diary' or 'register' which monitors the dates and contractual 'shelf life' of all early warning notices and compensation events and responses/lack of responses/disagreements, etc.

If the Contractor fails to respond in the prescribed manner and timescale, this needs to be followed up immediately. Perhaps a regular fortnightly meeting is the answer, in which the parties notify and discuss the relevant matters in a structured and time-governed manner. This also has the benefit of taking some of the heat out of contentious matters.

Administrative requirements

To operate the procedures will inevitably require additional technical staff. You cannot operate the NEC on a shoestring!

Changes to the subcontract

NEC3 stipulates that no change to the subcontract has effect unless confirmed in writing and signed by both parties. So Subcontractors need to be careful about lapsing into 'ad hoc' procedures on site, unless formally agreed as an amendment to the standard rules.

Don't forget the time effects

The NEC contract is clear. Quotations for Compensation Events must include cost and time effects. And once a quotation is accepted, it can't be reopened. One of the core principles of NEC is the contemporaneous assessment of compensation events (as opposed to the 'final account wrap up' approach of other standard forms).

This means that Subcontractors who don't include the time effects (in terms of cost or time) in their quotations are forgoing the opportunity to recover these costs.

If prolongation or disruption costs come to light after a quotation is accepted, e.g. direct costs, prelims costs and delay damages, they cannot be recovered under the contract. So, it's essential to consider time effects on every single Compensation Event.

Conclusions

All in all, I would advise you to approach and conduct the NEC contract with

- Very careful preparation
- Appropriate tender allowances for conformance
- An enlarged job staff
- A proactive attitude which reflects the NEC philosophy
- Pre-contract training of all technical, supervisory and administrative personnel
- Diligent adherence to the rules and time limits

If the correct approach is adopted, and the rules closely adhered to, then the NEC contract has a great deal to offer all concerned. However, a careless, under-resourced and/or unprepared approach will very likely lead to disaster.

Follow the NEC rules or risk the consequences!

Section 9

Real life problems

Brave new world
or con trick?

In the Third Edition Jack said, '*I notice an increasing gap between the brave new world of the construction industry's "chattering classes," and real-life events on site. One reads of a whole new approach, where "confrontation" is a thing of the past, and subbies are chosen on the basis of their proactive attitude. In this wonderful new world, the builder's door is always open, and the subbie is welcomed like a dear friend whenever he calls with a problem. "Claims" are a thing of the past. "Partnership" is the new buzzword*'.

Jack also said that the real world is somewhat different, and all this time later the real world is very different indeed! In fact, it is no exaggeration to say that, in my view at least, the construction industry is broken. It is dominated by large Contractors and contracts are let by them on onerous terms, in which the Construction Act is given a thorough mauling by the lawyers. The subbie is told not to concern himself with such trivia, but concentrate on 'being proactive' and that 'none of his competitors will bother'. Once on site, he finds the same old problems of late building and poor access to work faces, all symptoms of an overall delay in the project. If he raises these problems, he is accused of 'getting contractual' and reminded of other enquiries soon to be released for tender – but only to 'proactive' Subcontractors.

The subbie soon finds he is expected to jump from one work face to another, in an ad hoc manner. Programmes are abandoned. Milestone dates are 'written in stone', and he is expected to flood the job with labour and work weekends. At this point, the traditional threatening phone call at top level will be received – accompanied by threats of liquidated damages and set-off.

All this 'proactivity' tends to cost the subbie a great deal of money. An extension of time would be nice. Some reimbursement would be even better. However, our subbie is now referred to those onerous conditions. The Contractor has also changed the QS six times during the course of the job, and the latest 'hatchet man' doesn't recognise any of the variations and has scythed their value. Everything the subbie thought he had 'agreed' has been swept aside.

There are now demands for him to prove that he has followed the rules in every particular, with timely notices and records. No matter that he was originally told to ignore them! Individual 'cause and effect' must now be proven. He finds that the Contractor's door is no longer open. Communications are now handled by the 'Gestapo' (alias the Contractor's quantity surveying department), and it may take a very long time.

So, the moral is to be proactive, but to comply with the rules for notifying and recording delays and loss and/or expense. You must submit all the details required of you and get tough if you have to.

If you don't, you may end up a very unhappy subbie and one of the casualties of this brave new world. Don't believe it can happen to you?

Construction News 7 November 2017

Lakesmere's collapse a sad sign of the times:

> Lakesmere Group's collapse sent shockwaves through the industry this week in another hammer blow for the sector.
>
> This was the UK's largest building envelope specialist [approx. £120m turnover], according to CN's annual Specialists Index, and the biggest industry clients from Crossrail to Network Rail were scrambling to ensure projects didn't suffer this week.
>
> As seems so normal these days, problem projects were attributed as the principal factor behind the group's demise. You only need to look at the ongoing woes at Interserve and Carillion **to know there is something rotten with contracting** at present.

Don't be fooled, you must protect yourself in the 'brave new world'.

The real world – set your stall out from day one!

All those years ago, Jack referred to the 'love affair between the 'partners' at the top of our industry', and he said that back on the site little has changed. Boy oh boy, would he be sad to know that nothing much has changed. Except perhaps the amount of BS that some of the lovebirds spout! Indeed, for the subbie's site engineers and supervisors, it has now become much more difficult. If they dare to stand up for their firm, they are accused by the Contractor of being 'contractual', 'confrontational' or 'negative'. All too often, their own bosses fail to stand up for them, and they are 'hung out to dry'.

Let us begin by summarising the history of the typical construction subcontract:

1. The 'honeymoon' period.

2. Building not really 'fit' for M&E commencement.

3. Subbie works on 'seek and find' basis.

4. Building falls further behind.

5. Access delays and obstructions, building shell leaks like a sieve.

6. Variations flow through from the design team.

7. Contractor now in serious delay.

8. Contractor issues a revised programme. The works have been dramatically compressed, and the whole concept is absurdly optimistic.

9. Subbie is threatened with 'failure to use best endeavours'.

10. Subbie is coerced into increased labour, working weekends.

11. Project even further behind.

12. Subbie requests extension of time and loss and expense.

13. Contractor ignores him.

14. Client moves in before building gets completed.

15. Subbie must now complete works in occupied environment.

16. Client complains because of disgraceful unfinished state of the building services.

17. Contractor sets off damages for delay and contras for attendance, rubbish, etc.

18. Subbie again requests extension of time.

19. Contractor challenges subbie to produce notices, prove 'cause and effect' of delays and costs. Fetches in 'hired gun' to ensure subbie is denied his just entitlements.

20. Subbie now heavily 'in the red'.

21. Typical end of job dispute takes months or years to resolve.

22. At worst, subbie becomes insolvent and Contractor celebrates another successful project.

You may think that I am exaggerating but time and again, I see this very familiar pattern unfolding.

You would do yourselves a lot of good by **accepting that the above is typical, and setting out your stall from day one**, with the object of avoiding as much of the agony as possible and, above all, avoiding that end of job dispute. This means starting out with a fair set of terms, clear timescales, an agreed baseline programme, good site diary, regular progress reports, site photographs, no variations without instructions, etc., and making sure all delays and claims for recovery of loss and expense are formally registered in writing as and when they become apparent.

If you do this, you will be ready and able to respond to the various dramas as they unfold, and you will be able to protect your entitlements, thereby reducing the occurrence disputes and protecting your business and your sanity.

Set your stall out from day one and protect your entitlements!

What is normal?

'I don't know what you're complaining about' says the Contractor. 'It's just a "normal contract" – you must have worked on this kind of project before'. And in a sense, he is right. What passes for 'normal' on construction sites in recent years would be regarded as absolute madness by the proper builders of the past.

All manner of trades brought on to site before it is ready, having to wander around the building looking for work faces, falling over dozens of other subbies, only later to be told it is their problem to accelerate at their own cost in order to make up for the Contractor's delays.

However, let's take stock of the situation, and, in particular, consider what we have actually contracted to do. Did the enquiry documents tell you that you would be required to work in this manner? Do the standard subcontract forms (i.e. JCT) tell you to include for this kind of thing in your price? No, they don't. And if they had done so, then you would have put in a much higher figure, or maybe even declined to quote.

So, my advice is to stand your ground. If delay and disruption are 'normal' and, by inference, to be accommodated by the subbie at his own cost, why do the standard forms contain provisions for claiming loss and expense, revised rates, etc.?

It is no business of yours, when pricing the enquiry, to include for the likelihood of delays and disturbances caused by the Client, Architect and/or Contractor, unless such circumstances are expressly set out in the enquiry documents. The fact that such events are very probable is merely a sad reflection on today's broken construction industry.

Many of the possible events, such as failure to provide timely access and information, are in essence 'breaches' of contract, for which the subcontract conditions provide specific remedies (i.e. 'loss and expense').

So, my advice is to ask yourself this simple question 'Is this what the estimator priced to do?' Viewed from this starting point, it is obvious that the estimator had no reason to include in his pricing for delays and disruptions. When the streetwise subbie encounters, or even foresees, delays and/or disruptions, he should notify the Contractor in writing, and stand his ground in a calm and professional manner.

And to keep track of all this from the enquiry stage onwards, I would recommend the introduction of RAID logs into your systems. The acronym RAID stands for Risks,

Assumptions, Issues and Dependencies. It is a good idea to create a RAID log at the tender stage, so you can track anything that is going to impact you now or in the future.

Risks – Events that will have an adverse impact on your project if they occur. Risk refers to the likelihood of it occurring and the impact on the project if it does occur. If the likelihood of it happening and impact to the project are both high, you identify the event as a risk.

Assumptions – Any factors that you are assuming to be in place that will contribute to the successful result of your project. The log includes details of the assumption, the reason it is assumed, and the action needed to confirm whether the assumption is valid.

Issues – Something that is going wrong, or could go wrong, and needs managing. Failure to manage issues may result in a poor delivery or even complete failure. The log includes descriptions of each issue, its impact, its seriousness and actions needed to remove it.

Dependencies – Any event or work that is either dependent on the result of your project or on which your project will be dependent. The log captures what you are dependent on, what should be done and when. It may also include who is dependent on you.

You can create a simple RAID Log using a spreadsheet with one tab for each area. Use it as a primary source for your status reporting. Using a RAID log is easier than trying to keep all this information in your head and will pay enormous dividends.

Of course, you can always ignore this advice and succumb to the madness – you just won't ever make any money!

Ask yourself 'Is this what the estimator priced to do?'

Provisional sums

Unfortunately, 'provisional sums' can be a source of problems and disputes for the unwary subbie. A provisional sum is an allowance or best guess, inserted into enquiry documents for a specific element of the works that is not yet defined in enough detail for tenderers to accurately price. However, some subbies will also include their own provisional sums in their tender.

And it is the last sentence of the extract from Lord Justice May's judgment below which identifies where the problem lies, and why you need to be very careful indeed. Because, provisional sums are provided for in different ways in different forms of contract, and some forms of contract can be a little vague about how provisional sums should be handled, particularly regarding adjustments to the programme and associated preliminaries.

As provisional sums are replaced by valuations of the work actually done, the contract sum may increase or decrease. In addition, the actual works that are undertaken may affect your programming, planning and recovery of preliminaries. It is important therefore that you check the exact wording and meaning of the relevant clauses that are being used otherwise the innocent use of a supposedly familiar term could prove very costly indeed.

> In Midland Expressway Ltd v Carillion Construction Ltd Lord Justice May suggested that
>
> > The term 'provisional sum' is generally well understood in the construction industry. It is used in pricing construction contracts to refer either to work which is truly provisional, in the sense that it may or may not be carried out at all, or to work whose content is undefined, so that the parties decide not to try to price it accurately when they enter into their contract.
>
> The contract usually provides expressly how it is to be dealt with.

Don't be caught out by the use of provisional sums.

Dirty negotiating tricks and what you can do about them

Learning to negotiate

Don't let the prospect of learning to negotiate fill you with horror.

I'm not talking about how to handle hostage situations or persuade others about your political point of view, or haggle for a bargain in a Turkish street market!

What I am suggesting is that you take an objective look back at the situations you have been in where you would have benefited from a little bit of negotiating skill. How many times have you settled a final account for less than you were asking for, and probably less than it was worth? How did you handle that negotiation? Could you have done it better?

I do not consider myself to be an expert negotiator, but I am probably better than an awful lot of the people I come across in business. I don't mean to be rude to anyone and these same people are probably better golfers, engineers, photographers, computer buffs, footballers, etc. than I will ever be.

Negotiating is a skill that can be learned, and it is a skill which improves with practice; an excellent work on the subject is *Getting to Yes, Negotiating an Agreement without Giving In* by Roger Fisher and William Ury and for the second edition, Bruce Patton.

This great little book will introduce you to the following concepts:

- Separate the people from the problem.
- Focus on interests not positions.
- Invent options for mutual gain.
- Insist on using objective criteria.

Like most subjects, there is an immense body of information about negotiating, but these are some of the key concepts. For my sins, I have qualified as a CEDR Accredited Mediator, and this little book is recommended reading for what is a very intense but enjoyable course.

Separate the people from the problem

It is sometimes easy to forget that you are dealing with human beings. As the book says 'They have emotions, deeply held values, and different backgrounds and viewpoints; and they are unpredictable. So are you'.

This human aspect can be a good thing or a bad thing. Sometimes trust, understanding, respect and even friendship can help to resolve matters, whilst occasionally anger, frustration, hostility and egos can make life very difficult. Another very real problem is that human beings very often confuse their own personal perception with the objective reality of the situation.

Separating the people from the problem is all about recognising this human interest and literally dealing with it as a separate issue to the main issues about which we are negotiating.

Focus on interests not positions

In every negotiation, you need to look behind the other side's position and to establish their underlying interests. The example the book gives is of two men arguing in a library about whether the window should be open or closed. The astute librarian asks why one wants the window open: 'To get some fresh air'. She asks the other why he wants it closed: 'To avoid the draught'. Her solution is to open the window in the next room thus bringing in fresh air without any draught.

Establishing underlying interests is easier said than done, but it can be done. Each party should be willing to establish their underlying interests and to communicate them to the other side. Explain your interests to the other side and be prepared to acknowledge theirs.

If each party can think about their interests and consider possible solutions with an open mind, the chances of a 'win-win' solution will be increased dramatically.

Invent options for mutual gain

This is not as crazy as it may at first appear, but it is a difficult concept to explain briefly.

Sometimes, each party will see their task as narrowing the gap between their respective positions or trying to find the one single answer. Inventing new options may not come very easily for either party; but by looking for the single best answer, we may overlook a better solution selected from a wider range of options.

Insist on objective criteria

The basic concept at work here is that the discussions should include or be based upon objective criteria and not each party's perception.

Objective criteria should be independent of each side's will and should be both legitimate and practical. As the book says

> A principled negotiator is open to reasoned persuasion on the merits; a positional bargainer is not. It is the combination of openness to reason with insistence on a solution based upon objective criteria that makes principled negotiation so persuasive and so effective at getting the other side to play.

Unfortunately, not everyone you will come into contact with is a 'principled negotiator', so let's have a look at some dirty negotiating tactics and what you can do to counter them.

Dirty trick	Phoney facts
Description	Typically, the numbers appear valid, but the assumptions upon which the numbers or conclusions are based are dubious.
Example	'You have agreed to pay for 3 weeks of delay costs, and they come to £5,000 per week plus overheads'.
Counter tactic	Ask them to state their assumptions or explain how they derived their numbers.

Dirty trick	Higher authority
Description	It is common for a negotiator to delay reaching an agreement by claiming that his or her authority is limited. The other party will become impatient and give in to the earlier demands. Alternatively, it gives the negotiator an out if the settlement is not desirable.
Example	'Well, your proposal sounds interesting, but I will have to take it back to my boss for final approval'.
Counter tactic	Find out who the person with authority is before negotiations begin and only negotiate with that person.

Dirty trick	Intimidation
Description	A variety of influence tactics fall into this category including anger, fear, emotional ploys and guilt. They may also claim there are legitimate channels to go through.
Example	'If you don't accept our offer we won't pay you anything!'
Counter tactic	Remember that you will regret being bullied into a bad deal from which there is no escape. Keep calm and do not be bullied!

Dirty trick	Take it or leave it
Description	This is really not a negotiation approach. It is, however, an approach to conducting business which blocks negotiating.
Example	'This is all we are prepared to offer. If it is unacceptable, then you can take us to court!'
Counter tactic	Call their bluff. Ignore it and keep on negotiating!

Dirty trick	High ball/low ball
Description	Negotiator starts with an extremely high or low opening offer. Risk is that other party may consider negotiation is a waste of time.
Example	'We will settle your account at £100,000 (when it is worth £250.000)'
Counter tactic	Have objective information to counter their offer. Ask them to justify the offer and/or how they arrived at it.

Dirty trick	Chicken
Description	Negotiator combines a bluff with a purported action. It is a high risk strategy. If the other side calls their bluff, they must be willing to carry through with the action.
Example	'If you don't pay our claim we will take you to adjudication!'
Counter tactic	Ignore the bluff and keep on talking or call their bluff and bring the discussion to an end. But be sure of your ground before you do!

Dirty trick	Threats
Description	A useful tactic if one party has the power to inflict relatively large punishment on the other without substantial retaliation. The threat must be believed and you must be willing to follow through with the threat.
Example	'If that area is not completed by Friday, I will terminate your contract!'
Counter tactic	Make the threat salient and discuss the ramifications. Know your entitlements under the contract, and what they would be obliged to do before taking such action. If need be, review the contract terms together.

Don't fall prey to dirty negotiating tactics!

Get it in the budget

Put yourself in the shoes of the Contractor for a minute. You have financial control of a small army of subbies, of all shapes and sizes. You have to keep track of the budget and each month you fill in a 'liability statement', in which you set out what you anticipate the final costs are going to be.

If a subbie has notified and presented his claims in a professional way, and showed every intention of determined pursuit, then you're probably going to make an allowance for them, regardless of the finer points of contract law, etc. If, on the other hand, a subbie has done little or nothing in that direction, and showed every sign of being a 'push over', then you won't make an allowance.

Note that this approach has nothing to do with things like justice or fair play, it is quite simply a pragmatic judgement based upon perception of what you are going to have to pay.

It follows, therefore, that if a subbie who has made little or no noises during the job comes in with a glossy claim document after the works are finished, you have a problem, and frankly so does he. Because there is no money in the tin to pay him, even if his claims are rock solid.

I have every reason to believe that Contractors handle their budgets in such a simplistic manner. So, for heaven's sake, if you think you have a claim, or are likely to have one, tell the Contractor at the time. Provide reasonable estimates of loss and expense. Price your variations as you go along, not at the end. Make sure that all these figures are reflected in your monthly application. And, above all, apply plenty of pressure in writing and at meetings. If you 'get it in the budget', then you have a much better prospect of success!

Give early warning of all variations and claims.

Stand by your man

The late Tammy Wynette, a great country star of the 1970s, had a big hit with a song called 'Stand by your man'. I sometimes wish some of our director readers had been fans. Perhaps the words might have stuck, the way that lyrics do.

Contractors have a sure-fire way of dealing with awkward subbies. By 'awkward' I mean site staff who aren't prepared to let the Contractor walk all over them, but stand up for their firm's rights. The project manager will telephone the subbies' boss and accuse the site staff of obscure crimes such as being 'confrontational' or 'uncooperative'. Veiled hints would be dropped regarding 'future tender lists'.

Almost without exception, the reaction of the subbie's boss will be to rush out to site, admonish or replace his site staff, without establishing any of the facts, and generally put on a display of grovelling. Little does he know how the Contractor will fall about in hysterics after he has departed. No more trouble from that quarter!

How absurd, really, that the random mixture of abuse and half-truths should be given credence against men who have served for years with proven competence and loyalty. And yet, that is so often what happens. Please realise, that this is playing into the hands of those unscrupulous Contractors who use this sort of ploy as a standard management technique.

My advice, therefore, is to stay cool, investigate the allegations in an objective manner and to respond only when the full facts have been established.

And please remember Tammy's advice and 'Stand by your man!'

If they are right, give your site staff total support.

Business 'partnerships' are not love affairs!

In the third edition, Jack referred to the relationship between subbie and Contractor as like that between 'fair maiden and ardent suitor'. In the early days of courtship, the young man will turn up, always on time, often bearing flowers or chocolates. However, as time progresses the supply of flowers and chocolates dries up, and the young lady finds herself increasingly neglected.

Irrelevant nonsense? Not at all! Just like Jack, I have seen 'love affairs' between subbie and Contractor that have gone wrong. Whenever a Client says something is 'not a problem' because of the relationship enjoyed with the Contractor, I have to bite my lip. Now don't get me wrong – it is a great benefit to all when there is a good working relationship, where problems can be raised and discussed as they become apparent, and the works conducted in an atmosphere of mutual respect. However, too many subbies are conned into believing that this means they can ignore their contractual obligations to issue delay notices or formal correspondence. No wonder the Contractor loves them!

If things start to go wrong, the Contractor will murmur soothing words about 'seeing you right on the next job'. And so, our gullible subbie doubles his labour force and works weekends without any prospect of recovery And worse still, the next job doesn't come.

How sad, but how predictable! So, my advice is to forge a good working relationship with the Contractor and other key parties, but maintain your normal processes of contractual protection (i.e. delay notices, progress records, etc.). After all, that is what the Contractor's subcontract conditions say that you must do.

Forge good relationships, but don't neglect your contractual notices and records!

Conclusion

In the Third Edition, Jack said that he had kept this little book simple and down to earth, and hopefully in bringing it up to date, I have not changed that approach. Nothing has happened in the last 10 years to change my belief that most of our disasters arise from simple, down to earth events.

Jack also said, *'If anything, the role of the trade and subcontractor has become even more difficult since the original publication'* and unfortunately, the situation has continued to deteriorate!

If ever there was a time for clear and unambiguous leadership in the construction industry, it is now. The tragic events at **Grenfell Tower and the liquidation of Carillion (the UK's second largest Contractor at the time)** in 2017 have very publicly highlighted what happens when we get construction wrong, but what we don't see is all the things that are wrong with construction in the United Kingdom but are hidden away under the surface.

No amount of hard work on a building's external decoration will overcome problems lurking under the surface. If the foundations are defective, no amount of papering over the cracks will resolve the real problems.

But sadly, that is what I see in a construction industry that is almost unrecognisable from the one I joined 48 years ago! Worse still, we seem to have developed a culture of spin, and BS, and smoke and mirrors that is not only infuriating but also actually exacerbating the problem.

The smoke and mirrors

Whilst the industry is facing a whole host of issues, payment is far and away the biggest and most obfuscated issue. But the simple truth is that if you don't get paid, you don't have a business.

If you spend most of your time trying to get paid, you have less time to address all the other issues that need your attention, and as for planning for the future, you probably can't think beyond the end of the month. So, why are we as an industry failing to recognise this and why isn't it being properly addressed?

Regrettably, it is such a major problem and such a controversial issue that even when it is discussed, the truth seems to be largely absent.

Here are some facts

Streetwisesubbie.com recently supported a piece of statistically significant academic research. With some 502 respondents, its findings cannot be ignored. Some 354 of the respondents were Subcontractors, and over 70% of respondents had a turnover exceeding £1 million. Proper firms facing very real problems. Here's what the survey uncovered:

- A total of 87.17% of Subcontractor's main issue of dispute was regarding payment, compared to 64.21% of contractors suggesting Contractors receives payment within the agreed terms more frequently than their SubContractors.
- Even on public sector projects, only 25% of Subcontractors were paid on time, and on private sector projects that fell to just 15%.
- Subcontractors were subject to set off on 88.7% of their projects, and between 9% and 34% (depending on turnover) said they were subject to set off on every project!

So, what's to be done?

A staggering 93% believed that neither the government nor the trade bodies were doing enough to tackle the problems in the industry!

So, I say let's start by cutting out the spin, and the BS, and the smoke and mirrors!

- Lots of Contractors don't pay their Subcontractors properly. What's difficult about saying that?
- Everyone knows what good payment looks like! Pay on time and value fairly.
- The Supply Chain Payment Charter isn't even legally enforceable, never mind the fact that only a handful of major Contractors have even signed up to it!
- The Duty to Report on Payment Practices legislation (requiring large companies to report on their payment practices) is so woolly that it won't tell anyone anything.
- Those companies that engage in such practices certainly won't be reporting on their under-certification and wrongful set off.

Why does it matter?

Here are a few recent headlines (2017) to contemplate:

- **'Suppliers and subcontractors are owed around £2bn by Carillion'**
- 'Subbies and suppliers owed £7bn in unpaid invoices'.
- 'Subbies owed £4m after Titan Construction Management collapse'.
- 'Subbies under attack from new breed of aggressive QS'.

And some people wonder why there's a skills shortage! Frankly, for an awful lot of Subbies getting paid is a one-sided battle, which they are never going to win without the big guns of proper legislation.

Without it we will never deliver the massive culture shift that is required to rid this industry of its morally bankrupt payment practices. And, as for all sitting round the table together, it isn't currently working, and I can't see it ever working. After all the 'Anti Fox Hunting / Pro Fox hunting Association' doesn't sound like a group that's going to please any of its members or agree on anything!

Whilst the government and certain organisations talk of 'progress' a 'world class industry' and a 'strong and sustainable supply chain', the reality is that nothing much has changed for the better, but plenty has changed for the worse!

Unfortunately, if the job goes pear shaped, the subbie is stumped for lack of evidence. At that point, the subbie finds that 'some partners are more equal than others'. Not that I want the subbie to go in on day one 'looking for trouble'. However, the simple fact is that virtually all subcontract conditions require the subbie to give early written notice of delay and additional costs. To shirk this duty does nobody any favours. Nobody likes nasty surprises at the end of the job.

By bringing possible problems to the fore at an early stage, the streetwise subbie will be giving other parties an opportunity to take action geared to avoiding or mitigating those problems. Some Contractors and/or Clients fail to see it in that light, and react in a defensive, even aggressive way.

The streetwise subbie must get it right from the outset; understand the contract; negotiate the best possible terms; do what the contract says; and above all, stay calm and professional, even when others are doing neither.

Appendices

Stand by your man!

Appendix 1

The enquiry to contract process

Use the flow diagram as a reminder that orders are a process not an event!

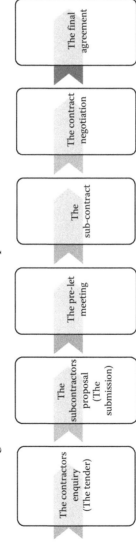

The contractors enquiry (The tender) → The subcontractors proposal (The submission) → The pre-let meeting → The sub-contract → The contract negotiation → The final agreement

Raid log page 1 of 2

Risks

Risk ID	Risk grouping	Risk Description	Risk Owner	Raised by	Date Raised	Likelihood	Impact	Score	Risk Strategy	Management action(s) identified	Target / Review date	Date Resolved	Status
R-001	Procurement	Fitting Supplies	Contract Manager	Installation Supervisor	01.01.2017	4	5	20	REDUCE	Current supplier, specific fixings	07.01.2017		Open
R-002	Needs a Plan	Protection of the works	Installation Supervisor	Contract Manager	01.01.2017	1	5	5	AVOID	Clause in contract stipulates protection method and duration, Ensure adequate measures are in place to comply	07.01.2017		As yet unplanned
R-003	MD Approve LADs	MD Approve LADs	MD	Contract Manager	01.01.2017	5	3	15	ACCEPT	LADs negotiated to a figure proportionate to the scope of works	07.01.2017	14.01.2017	Closed

Assumptions

Assumption ID	Date Raised	Raised By	Description	Validated By	Date of validation	Status
A-001	01.01.2017	Estimator	Drawing time for design is unlimited			Awaiting Design Managers opinion
A-002	01.01.2017	Contract Manager	The propsal was based on the Contractors requirements	The Contract	Date SWS reviewed Contract	Closed

Raid log page 2 of 2

Issues

Issue ID	Associated Risk	Issue Description	Issue Owner	Raised by	Date Raised	Severity	Likely Impact on	Management action	Action taker	Target date	Date Resolv	Status
I-001	R-002	Protection of works material is on an 8 week lead time, PC is in 2 weeks	Purchasing	Contract Manager	02.01.2017	Critical	LADs	Give notice & request substitute protection	Notice issued	What does the contract say	03.01.2017	COMPLETE
I-002	No risk identified prior to issue	Cutting up the internal crane in the building has caused iron filings to oxidise to aluminium frame of solar panel	Site Supervisor	Project Manager	02.04.17	High	Delay in handing over defect free	Cost variation for additional works to clean the panels	Notice issued	What does the contract say		OPEN

Dependencies

Dependency ID	Date Raised	Raised By	Dependency Description	Internal or External	Owner	What is dependant on delivery	Priority (1 to 5)	Actions / Next Steps	Next Review Date	Status	Date Closed
Dep-001	01.01.2017	Finance	Customer Payment History	Internal	MD	Payments for materials to achieve Pratical Completion	1	Call StreetwiseSubbie	07.01.2017	Red	
Dep-002		Contract Manager	Sequence of Works	Internal	Project Manager	Banksman & crane lifting plan required	5	Appoint 3rd parties	07.01.2017	Amber	
Dep-003											
Dep-004											
Dep-005											

Appendix 3

The streetwise subbie's site checklist

Here is just one idea for a site checklist, it is an aid to good contractual management and it can be adapted to suit your own particular circumstances. Any answers in the 'No' column are indicative of a potential or actual problem, so further investigation or action is almost certainly required.

Programme	Yes	No
Have start and finish dates and a programme been agreed?		
Reference/date original programme		
Original subcontract completion date		
Latest revised programme reference/date		
Current programmed subcontract completion date		
Has copy of Main Contract programme been obtained?		
Is actual subcontract completion date recorded and agreed?		
Is a programme register maintained up to date?		
Progress/completion/extension of time	**Yes**	**No**
Was actual subcontract site start date agreed?		
Was site start date as per agreed programme?		
Is progress as original programme?		
Is subcontract programme being compressed in order to overcome previous delays?		
If so, has agreement to extra costs been requested?		
Have delay notices been submitted to cover all delays to date?		
Receipt of information	**Yes**	**No**
Has all information been applied for, stating dates required?		
Has/was all information received as requested?		

	Yes	No
If not, have delays and effects been notified?		
Are variations excessive/disruptive/late?		
Has appropriate notice been given?		
Financial claims/variations	**Yes**	**No**
Are claims for delay costs notified?		
Are claims for disruption costs notified?		
Are variations submitted up to date?		
Has compression/acceleration been imposed/agreed?		
Have compression/acceleration costs been notified and agreed?		
Records	**Yes**	**No**
Is site diary detailed/up to date?		
Are dayworks detailed/up to date and submitted?		
Is labour register up to date?		
Is labour booked to programme activities?		
Are detailed weekly plant and tools records kept?		
Are weekly programme/progress percentages recorded?		
Is main contract progress being monitored?		
Are regular site meetings being held?		
Are minutes being received?		
Are minutes being corrected where inaccurate?		
Are/were all records maintained right up to present date and/or actual date of practical completion?		

Appendix 4

The monthly check-up

Here is an idea for expanding the site checklist into a monthly management checklist, it is an aid to good contractual management and it can be adapted to suit your own particular circumstances. Any answers in the 'No' column are indicative of a potential or actual problem, so further investigation or action is almost certainly required.

Project title:

Order/subcontract documents	Yes	No
Has official order been received?		
Date and reference of official order		
Has subcontract agreement been received?		
Date and reference of subcontract agreement		
Were terms and conditions agreed?		
If not, have objections been recorded?		
Has schedule of rates been submitted?		
Has schedule of rates been agreed?		
Programme	Yes	No
Has an original programme been agreed?		
Reference/date original programme		
Is this recorded in the order/agreement?		
Original subcontract completion date		
Latest revised programme ref/date		
Current programmed subcontract completion date		
Current realistic forecast subcontract completion date		
Has a copy of original Main Contract programme been obtained?		
Reference date of original Main Contract programme		

	Yes	No
Actual subcontract completion date achieved		
Is actual subcontract completion date recorded and agreed?		
Is a programme register maintained and up to date?		
Progress/completion/extension of time	Yes	No
Was actual subcontract site start date agreed?		
Actual subcontract site start date		
Was site start date as per agreed programme?		
Is progress as original programme?		
Is subcontract programme being compressed in order to overcome previous delays?		
If so, has agreement to extra costs been requested?		
Have delay notices been submitted to cover all delays?		
Has extension of time been formally requested?		
Has extension of time been granted?		
Current extended completion date (if any)		
Receipt of information		
Has all information been applied for, stating dates required?		
Has/was all information received as requested?		
If not, have delays and effects been notified?		
Are variations excessive/disruptive/late?		
Has appropriate notice been given?		
Financial claims/variations		
Are claims for delay costs notified?		
Are claims for delay costs submitted?		
Are claims for disruption costs notified?		
Are claims for disruption costs submitted?		
Claim sums submitted (delay)		
Claim sums (delay) agreed by Client/Contractor		
Claim sums submitted (disruption)		
Claim sums (disruption) agreed by Client/Contractor		
Is the variations payment applied for up to date?		
Are variations submitted to date?		
Are variations agreed by Client/Contractor?		
Has 'change of character/conditions' been suffered?		
Has notice and revised pricing been submitted?		
Is non-varied work affected by variations?		
Has notice and revised pricing been submitted?		

Has compression/acceleration been imposed?		
Has compression/acceleration quotation been submitted?		
Compression/acceleration sums submitted		
Compression/acceleration sums agreed by Client/Contractor		
Have set-off (contra charges) been attempted by Contractor?		
Have they been challenged?		
Records		
Is off-site management and/or staff time booked and costed to specific subcontract?		
Is site staff and supervision, including working chargehands, storekeepers, messing attendants, etc. all booked and cost to specific subcontract?		
Is site diary detailed/up to date?		
Are delay notices submitted and up to date?		
Are dayworks submitted and up to date?		
Is labour register up to date?		
Is labour booked to programme activities?		
Are labour time sheets maintained?		
Are detailed weekly plant and tools records kept?		
Are weekly programme/progress percentages recorded?		
Is Main Contract progress being monitored?		
Are regular site meetings being held?		
Are minutes being received?		
Are minutes being corrected where inaccurate?		
Are/were all records maintained right up to present date and/or actual date of practical completion?		
Are/were all notices maintained right up to present date and/or actual date of practical completion?		

Appendix 5

Good and bad correspondence/notices

Although giving notice is perhaps unfamiliar territory for lots of Subcontractors, it is something that MUST be done in order to protect your interests when things start to go wrong. Not only is it unfamiliar territory, it is also something that the Contractor will probably try and dissuade you from doing! DO NOT be put off.

So perhaps it's no surprise just how poor the letters and notices submitted by Subcontractors can actually be. What seems very clear and appropriate at the time when we are under pressure to get the job done can prove almost useless when trying to prove your case for delay at the end of a problem contract.

Some 'not so good as it seems' examples!

'We confirm our visit to site today and our view that the site is not ready for commencement of our works. We shall keep in close contact and visit again next week.' (*Why is it unready? Be specific!*)

'We refer to our conversation on site yesterday, and trust that you are now clear regarding the reasons for our delay in commencement of Level 2, which result from the default of others.' (*What are the reasons? Delay in commencing what activities? Who are 'others'?*)

'We confirm our site discussions yesterday, when we explained that we are unable to commence our installations on Level 2, due to lack of readiness by other trades. This may affect our programme.' (*What exact installations? What other trades? What is not ready? Which programme activity may be affected?*)

Angry/sarcastic/contractual correspondence

In some quarters, it was considered great stuff to send letters or emails full of sarcasm and semi-abuse. Apart from anything else, it might have made some people feel better. Equally, some practitioners teach that no delay notice is complete without a mass of clause numbers and 'barrack room lawyer' material. No wonder we had so many stand-up confrontations with builders and Clients! These days, the emphasis is on **protecting the company without losing the Client!** That means we have to be a lot more subtle in the way we write our letters and memos.

Some old-fashioned examples

'We have received your letter dated 8 January 2018 accusing us of delay on Level 1. We are amazed at your allegations, which just prove that you are completely out of touch with the real situation on site. This is another case where our willingness to be helpful has been used against us. If certain parties spent more time out on site and less time criticising others, then the job would be a lot further on etc.'. (*This is a wonderful example of the old-fashioned approach. Full of irrelevant emotion, and totally silent on facts! The Contractor must have gone through the roof when he read it. The fact that the subbie's remarks were probably true would have probably rubbed salt into the wound! And the Contractor would be getting his own back by knocking down our valuation and final account, and hitting us with contra charges.*)

'We have received your Programme 21B dated 5 January 2018. We are amazed to see that you still require us to finish on the original end date, regardless of the delays we have suffered. You have been notified on several occasions under Clause 20(1)(a) of delays due to late access, and also 20(l)(b)(i) and (ii) regarding late information. We have requested an extension of time under Clause 20(2)(a) and 20(3)(c)(iii). We are not prepared to work to your programme, which fails to take account of our previous requests. If you insist that we do, then we give you notice pursuant to Clause 26(3)(a) of our intention to claim all resulting loss and expense, including finance costs'. (*If all else has failed then OK. But otherwise, this example is too heavy, too 'legal' and looks like a subbie spoiling for a fight. There is no sign of any willingness whatever to help the situation. To a Contractor with his back to the wall, this letter would go down 'like a lead balloon'.*)

The same letters using a more subtle approach

Let us look at the same letters using a more subtle approach, designed to protect ourselves contractually without being too confrontational.

'We confirm our site visit on 12 February 2018 and thank you for taking the time to look around the site with us. The present state of building progress is not quite ready for a start on electrical works (i.e. roof incomplete, and considerable waterlogging in the basement, where we are due to commence first fix). We confirm our agreement that we will visit again in a week's time, and that we shall jointly review the situation. Assuring you of our full co-operation'. (*This example is factual, polite and non-aggressive.*)

'Thank you for giving us your time on site yesterday, when we discussed the progress situation on Level 2 first fix. As we explained, the reason for being unable to commence work in this area has been due to the general lack of weather-proofing, as a result of the general lack of progress of the roofing works. However, we are pleased to see that the situation is rapidly improving, and hope to move into the area with a squad of ten electricians on Monday 26 February 2018. Assuring you of our commitment'. (*This example gives details of the cause and exact activity affected, in a friendly and reassuring way.*)

'We are in receipt of your letter dated 8 January 2018 regarding progress on Level 1. We would respectfully refer you to our Delay Notices Nos 1267 and 1268 dated 4 January 2018 and 5 January 2018, when we reported that we were at a standstill on final fix work due to the whole area being occupied by floor layers. However, we have kept close contact, and note that the floor layers have almost finished their work. We shall therefore return with a full squad on Monday 15 January 2018, to expedite completion of the final fix to this area. Assuring you of our best attentions'. (*This example refutes the allegations, in a calm, factual and non-emotional way.*)

'Thank you for your letter dated 8 January 2018 enclosing your revised Programme No 21B dated 5 January 2018. Having studied same very carefully, may we respectfully comment as follows:

a) We note that our works are programmed for completion on the original end date of 31 March 2018.

b) Whilst we are anxious to assist in every way, we must refer to our various delay notices and requests for an extension of time of 6 weeks.

c) It is our view that completion by the original date could only be achieved by 'special measures' (e.g. additional labour and/or weekend working).

We are very willing to discuss this matter with you, in the interests of the project, and suggest a meeting early next week. Assuring you etc.'

(*This example makes the subbie's position clear, but shows a willingness to help in overcoming the Contractor's problems. There is no specific reference to how these 'special measures' are to be reimbursed, but the message is there for the Contractor – the message being 'We are prepared to accelerate, but only once an agreement has been reached'*).

Essential points of a 'good' letter or notice

It is not necessary to be a 'barrack room lawyer' or a 'smart Alec' in order to write a good letter. Some key points are as follows:

1. Put the subject heading at the top of every letter.

2. Make the letter 'self-contained', so a stranger can understand it in a year's time.

3. Confine the contents to the simple facts without emotion.

4. Don't delay – write the letter while the problem is happening.

5. Make sure you comply with the requirements of the contract (if it says a notice is only valid if sent by Special Delivery, then that is what you must do)

6. Make sure you send it to the right person/place (some Contractors include very specific requirements).

As to a good notice, the basics are as follows:

1. Put the notice in writing as soon as the problem becomes apparent.

2. Use an objective non-confrontational tone, but don't use 'victim language' ('*we have been prevented...*' rather than '*we are in delay...*').

3. State the specific problem causing the delay or disruption, the area affected, and the actions required from others.

4. Give a view as to the effects on programme and overall completion.

5. Request extension of time if necessary.

6. Give details of any obvious cost effects.

7. Update as necessary from time to time and record date when delay is cleared.

8. Show willingness to co-operate but not at our own expense

Be proactive!

Serve your notices, but don't get bogged down in a 'letter war' for its own sake. Show a willingness to talk to the Client and/or Contractor and explain your problem to them, discuss and propose possible solutions.

One often overlooked purpose of giving notice is to ensure the Contractor knows exactly what is going on, and to give him the opportunity to do something about the matters which are delaying you.

Obligations to accelerate

In the absence of an express provision within the contract, there is no general obligation on the Contractor or you the Subcontractor to accelerate the progress of the works where they have fallen behind programme because of matters beyond your control.

If the Contractor wishes you to accelerate, then he must secure your agreement. If he purports to issue a direction to accelerate, you can refuse to comply unless he can demonstrate that he is empowered to instruct you to do so under the contract.

So, watch out for non-standard clauses requiring free acceleration!

Back up the notice with evidence

Don't rely on anyone else to have details of the delay issue. They might be the party who ends up making the final decision if the issue gets referred to Adjudication or Arbitration, and/or be keen to see your claim fail.

Wherever possible, supply backup information to support the delay notice, if you can cross reference all other information, including, if possible, a programme showing the effect on the works. If you can't send backup information at the time, then follow it up as soon as you can with further information.

Appendix 6

Examples of site records

Site diary	Ref No 1230
CONTRACT: Royal Hospital	**DATE:** 8 April 2017
WEATHER: Fine all day	**VISITORS:** Mr G James, Mr L Jones (Head office) Mr B Able (architect)

PERSONNEL ON SITE

SUPVSRS 1 **FMN** 1 **C/HDS** 2 **APPVD ELS** 4 **ELS** 22 **APPCES** 4 **LABS** 4 **OTHERS** Driver 1, Site Clerk 1, Site Secretary 1.

SUB-SUBCONTRACTORS (E.G. CABLE GANGS AND FIRE ALARMS SPECIALISTS):
Smith Brothers (cablers) 1 Fmn, 6 operatives
High Tek (PA specialists) 1 Fmn, 2 operatives

AREAS/ACTIVITIES STARTED TODAY:
Block C Level 1 – 1st fix
Block A Level 6 Plant room – 1st fix
Block B Level 5 – 3rd fix

AREAS/ACTIVITIES COMPLETED TODAY:
Block A Level 4 – 1st fix
Block B Level 4 – 3rd fix

MAIN PROGRESS TODAY (other than starts/completions):
Block A – Levels 1 and 2 – 1st fix
Block B – Levels 1 and 2 – 2nd fix/final fix
Block E – Levels 1 and 2 – 2nd fix/final fix

VARIATIONS/INSTRUCTIONS RECEIVED TODAY:
SI No 56 receive from Contractor– Cancel Type A luminaires (advised too late to stop delivery)
for Block A (pending architect's review).

AI No 124 received from Contractor – Changes to lighting for Block G.

CURRENT DELAYS DUE TO ACCESS/BUILDING WORKS/OTHER TRADES:
Block G still unavailable – roof not completed
Block A Level 1 Rooms 45–54 still full of scaffolding – prevents 1st fix

CURRENT DELAYS DUE TO INFORMATION/DRAWINGS/VARIATIONS:
Still await architect's decision re X ray equipment for Block A.
Approvals still awaited re working drawings for Block B Sub-basement SW Room.

OTHER EVENTS: Received delivery of Type A luminaires – advised Contractor that variation will take account of restocking charges handling, etc.

SIGNED: R. Goodguy   **TITLE:** Site supervisor

Site delay report   Ref 2321

Contract: Royal Hospital   **Date:** 8 April 2017

To: Mr J Bloggs
Ace Builders Ltd

Re:
Block/Level/Area/Room: Block A Level 1 Rooms 45–54
Activity: Electrical 1st fix (Prog Activity A/24)

In accordance with Clause 2.17.1, we are obliged to give you notice of the following event, which is delaying our progress:
Rooms 45–54 still full of scaffolding, which continues to prevent us from starting Electrical 1st fix. Please see our notices dated 11/3/17, 18/3/17, 25/3/17 and 31/3/17.

Action/information required to enable progress:
Please can you clear for access urgently, as promised by your Mr Potter on 1/4/17.

Delay to programmed activity:
Prog. Activity A/24 shows commencement of Electrical 1st fix in Block A on 11/3/17 = 4 weeks prevention still ongoing.

Effect of delay:
Rooms 45, 46, 47 are main plant rooms and critical to electrical progress in Block A and project generally. Any delay will 'knock on' directly to overall programme completion date, all as previously notified.

Signed: R. Goodguy **Title:** Site supervisor

Technical query register

TQ No.	Date	Subject	Reply required	Reply received	Date resolved
1	1/2/06	SW rooms A 45/47 Panel positions	8/2/17	2/3/17	2/3/17
2	6/2/06	Block B Lighting scheme	13/2/17	20/2/17	See 2A
3	14/2/06	Block C Cable spec	21/2/17	21/2/17	21/2/17
2A	21/2/06	Block B Lighting scheme	28/2/17	7/3/17	See 2B
4	3/3/06	SW rooms wiring	10/3/17	10/3/17	10/3/17
2B	7/3/06	Block B Lighting scheme	9/3/17	22/3/17	See 2C
5	15/3/06	Block A Level 1 X ray equipment	22/3/17	23/3/17	See 5A
6	22/3/06	Fire alarms specification	29/3/17	25/3/17	See 6A
2C	23/3/06	Block B Lighting scheme	25/3/17	8/4/17	See 2D
5A	23/3/06	Block A Level 1 X ray equipment	24/3/17	26/3/17	See 5B
6A	25/3/06	Fire alarms specification	27/3/17	27/3/17	27/3/17
7	28/3/06	PA wiring generally	10/4/17		
5B	3/4/06	Block A Level 1 X ray equipment	5/4/17	5/4/17	5/4/17
8	7/4/06	Block H Level 3 Path Lab lighting	17/4/17	21/4/17	21/4/17
2D	8/4/06	Block B Lighting scheme	11/4/17	20/4/17	20/4/17
9	10/4/06	Block A, B, C Emergency lighting	17/4/17	15/4/17	15/4/17

Technical query   Date: 1/2/17   No. 1

To:
Mr J Bloggs
Ace Builders Ltd

We should be pleased if you would provide us with your clarification/instruction/ information regarding the following query, by the date requested:

Block: A   **Level:** 1   **Rooms:** 45/47
Drawing/s No: Red/1327/HAR/12 Rev A dated 26/1/17

Activity/detail: SW panel positions

Request: Please advise us of exact positions of SW panels shown on above drawing as 'to be finalised'.

Comments: This information is urgently required as we are currently programmed to be working on 1st/2nd fix electrical installations in these rooms, and therefore urgently need the precise positions, to avoid 'looping' of cables, etc., and unnecessary return visits.

Response required by: 5/2/17

Signed: R. Goodguy   **Title:** Supervisor

Response:

Attach herewith sketch ref SK RED/1327/57, which shows precise positions of the 3 No SW panels referred to, complete with dimensions. Also refer to site discussions between your Mr Ian Keen and our Mr B. Stubbin of today's date, and joint marking up of walls.

Signed: J. Bloggs Title: M&E Coordinator **Date:** 5/2/17

Site communication / email   **Ref.** 2734

To: Mr J Bloggs

Ace Builders Ltd

DATE: 10/3/17

:

Request for access/works by others:

(Delete which inapplicable)

Block: J Level: 1 Room: 42

Activity: Electrical first fix

We are programmed to start work in the above room on Monday 24/3/17. At present, the area is being used as a store for plastering materials. Please would you arrange for the room to be cleared in time for our programmed commencement.

Thank you for your assistance.

Signed: R. Goodguy **Title:** Supervisor **Date:** 10/3/17

Progress report No: 7 Dated: 5/5/017

Block	Level	Activity ref.	Activity	Prog. start	Actual start	Prog. finish	Actual finish	Prog. (%)	Actual (%)	Comments
A	1	12	Electrical 1st fix	3/3/17	17/3/17	24/3/17	–	100%	75%	Await access Rooms 45–56 incl.
A	1	13	Electrical 2nd fix	10/4/17	30/4/17	1/5/17	–	100%	33%	Ditto
A	1	14	Electrical 3rd fix	24/4/17	–	15/5/17	–	33%	0%	Ditto
A	2	15	Electrical 1st fix	20/4/17	10/4/17	10/5/17	–	60%	75%	Switched from Level 1
A	2	16	Electrical 2nd fix	3/5/17	29/4/17	24/5/17	–	5%	33%	Ditto
B	1	20	Electrical 1st fix	24/4/17	4/5/17	15/5/17	–	50%	0%	Await delayed approval of working dwgs submitted 10/3/06
C	1	28	Electrical 1st fix	10/3/17	–	24/3/17	–	100%	0%	Whole area on hold since 10/3/06 awaiting redesign of all services

Daily allocation sheet Day: Monday Date: 5/5/17							
Name	Trade	Block	Level	Prog. activity	Activity	Variations	Other/comments
R. Goodguy	Sup						General supervision
A. Stalwart	Fmn						General supervision
I. Steadfast	C/hd	B	3			AI No 25 – addnl heaters	50% supervisory
A. Trusty	C/hd	A	4	16	Electrical 1st fix		50% supervisory
B. Loxley	A/Elec	A	4	16	Ditto		
T. Lawton	Elec	A	4	16	Ditto		
T. Johnstone	Elec	A	4	16	Ditto		
E. Houghton	Elec	B	2	24	Electrical 2nd fix		
A. Southwell	Elec	C	1	32	Electrical 1st fix		
H. Brown	Elec	B	3			AI No 25 – addnl heaters	
B. Corkhill	Elec	B	2	24	Ditto		
E. Lowe	Elec	B	2	24	Ditto		
J. Sewell	Elec	B	3			AI No 24 – addnl hand dryers	
	Elec	J	1				External ltg – unprogrammed
B. Baxter	Elec	B	3			Ditto	
T. Deans	Ap Elec	B	3			Ditto	
N. Rigby	Ap Elec	B	3			Ditto	
J. Adamson	Ap Elec	A	4	16	Electrical 1st fix		
F. Broome	Lab	C	1	32	Electrical 1st fix		
W. Evans	Lab	B	2	24	Electrical 1st fix		

Programme register

Programme title	Prog. ref.	Prog. date	Issuer	Date issued/recvd	Status
Draft construction programme	Ace/tm/1	8/6/17	Ace Builders Ltd	10/8/17	For information
Master construction programme	Ace/tm/1A	3/9/17	Ace Builders Ltd	21/9/17	For construction
Draft programme – electrical	Sparks-em draft	30/9/17	Sparks Ltd	5/10/17	For comment
Electrical services programme	Sparks-em/2	14/10/17	Sparks Ltd	18/10/17	For approval
Electrical services programme	Sparks-em/3	22/10/17	Sparks Ltd	29/10/17	Agreed subcontract programme
Target construction programme	Ace/tm/1B	14/11/05	Ace Builders Ltd	24/11/05	Revised – for construction
Electrical services target programme⟶	Sparks-T1	14/12/05	Sparks Ltd	21/12/05	For comment

Progress photographs register

Photo ref	Photo date taken	Location	Comments
36	5/5/17	Block A Level 1 – Corridor to rooms 45/56 incl.	No access. Obstructed by rubble, scaffolding, loose materials
37	5/5/17	Block A Level 1 – Room 45	Being used as store for unfixed plaster boarding etc.
38	5/5/17	Block A Level 1 – Room 47	Occupied by scaffolders and blocklayers
39	5/5/17	Block A Level 1 – Room 56	Ditto
40	12/5/17	Block A Level 1 – Corridor to rooms 45/56 incl.	No access. Obstructed by block layers
41	12/5/17	Basement plant room	Widespread water ingress. Lack of internal walls
42	12/5/17	Admin block – Main staircase	Cascading with rainwater following overnight rain
43	12/5/17	Admin block – Main reception area	Windows incomplete. No protective sheeting. Whole area waterlogged
44	12/5/06	Site access road – viewed from Lancs Road	Quagmire following recent rain. Vehicular access impossible

Barry J Ashmore Dip Law Dip Arb FCIArb MCMi

Ashmore Consulting - Derbyshire

t: 01773 715062   m: 07815 193718   e: barry@ashmoreconsulting.com

Ashmore Consulting work exclusively with Specialist Contractors in the UK construction industry, and provide expertise in dispute resolution, contractual and commercial problem solving, contract evaluation and drafting, training seminars and business development.

Our expert team are professionally qualified and have over 40 years of construction experience, and 28 years' experience of helping Specialist Contractors in the construction sector to resolve disputes using our advanced negotiation skills, adjudication, litigation, mediation, and arbitration expertise.

We have helped Specialist Contractors of all sizes and specialisations to get paid and protect themselves against contractual and commercial issues, and resolved disputes from a few £thousand to several £million pounds.

We operate in all parts of the UK and overseas.

ashmoreconsulting.com

Experts in Construction Law for Specialist Contractors.

Barry J Ashmore is also the Managing Director of StreetwiseSubbie.com Ltd, the nationwide network of professionals and online advice service for Specialist Contractors. Free information and resources for Specialist and Trade Contractors can be found at;

Streetwisesubbie.com

Index

Ury, William, 159
User-friendly systems, 79

V
Valid applications, 122–123
Valuation, 128
  of variations, 73–74
Variations, 68
  change of character/conditions, 73–74
  true costs of engineering and supervision
    involvement, 75–77

W
Weekend guests, 57
Weekend working, 55
Weekly time-related actual costs, 88–90
Withholding notice, 109–110
Workshop, 5
Written notice, 81, 83, 89, 93, 101
Wynette, Tammy, 165

Z
Z clauses, 145–146